WHY BUY VEGETABLES?

SAVE MONEY BY GROWING YOUR OWN

WHY BUY VEGETABLES?

SAVE MONEY BY GROWING YOUR OWN

BY VEERE SHERREN

NEW ENGLISH LIBRARY
TIMES MIRROR

Contents

Copy Editor: William J. Howell
Art Editor: Deborah Miles

This edition published in 1978 by
New English Library Limited,
Barnard's Inn,
Holborn,
London EC1N 2JR,
England

Set in 10/12pt Monotype Univers by South Bucks Typesetters Limited
Printed by Fratelli Spada, Ciampino, Rome, Italy

450 03744 4

ACKNOWLEDGEMENTS
The publishers and editor of this book are greatly indebted for the many fine colour illustrations it contains to the Harry Smith Horticultural Collection ; Suttons Seeds Limited ; and Thompson & Morgan Ltd. Their addresses are :
Harry Smith Horticultural Collection, Hyde Hall, Rettendon, Chelmsford, Essex, CM3 5ET.
Suttons Seeds Limited, Hale Road, Torquay, Devon, TQ2 7QJ.
Thompson & Morgan Ltd, The Seedsmen, Ipswich.

Introduction
Why Buy Vegetables?

Let it be said from the very beginning that, in the main, growing vegetables is easy and rewarding in every sense. Growing your own is also wise because they taste much better from your own garden and can save considerable money, particularly if a deep freezer is available, when any surplus can be put down for eating at a later date. There is a great sense of personal achievement as well as a consuming interest and hobby for the gardener, but it must be remembered that *even the professional* has an occasional failure. Vegetables do not need to take up too much room in the garden if care is taken in selection, because so many crops follow one another seasonally. For instance, when a row of green peas is over in July, it is time to plant your winter cauliflower. When broad beans are finished, you can use that same piece of ground to plant late carrots, and so on.

Some useful hints

The secrets of success are very simple:

* Dig the ground well in October and November and leave it in large lumps for the frost to break up during the winter, thus avoiding the soil being packed down hard by winter rains.
* Give a generous amount of cow or horse manure at the time of digging, at least 8 in. (20 cm) under the soil.
* In the spring, give a good raking 3 in. (7.5 cm) deep or, better still, use a rotavator which breaks even the heaviest soil down into a fine top soil. A light dressing of dry manure and lime may also be added at this time.
* Feed the top soil with a proprietary general fertiliser, but never give more than the amount recommended by the manufacturer whose instructions are clearly indicated on the container, and also add peat to the top soil when raking.
* Water well when planting and in fine weather continue watering at weekly intervals. Unlike flowers, vegetables never fully recover from a severe check, in growth from lack of moisture. When you look at the percentage of water in various raw vegetables, you can understand how important water is to your vegetable crop:

Vegetable	% Water
Asparagus	91.7
Beans (French and Runner)	90.1
Beets (Roots)	87.3
Broccoli	89.1
Cabbage	92.4
Cauliflower	96.0
Celery	94.0
Chives	91.0
Cucumbers	95.1
Lettuce	95.0
Melon	92.6
Peas	78.0
Peppers	93.4
Potatoes	79.8
Radish	94.5
Tomatoes (Ripe)	93.5

* Hoe regularly to keep down weeds and loosen soil so that air, rain and water can go down to the roots.
* Spray with proprietary insecticides at regular intervals to prevent and reduce attacks from insects and diseases.
* Rotate crops on each piece of ground each year, so that no one piece of land grows the same crop in two consecutive years. (See Chart on page 14).

  Do not plant any one crop all at the same time. For instance: with runner beans, plant one third of the row and then three weeks later the next one third and then, finally, the last one third later still, thus increasing the length of the harvesting period. With many vegetables seeds for early, middle-period and late varieties, can be bought, which is a wise procedure to follow.
* A few fruits, such as strawberries, raspberries and gooseberries are dealt with because, although they are not vegetables, they are very much a part of the vegetable garden for most amateur gardeners.

Disease-resistant seeds

Undoubtedly, the very best way to avoid crop failure from disease is to select seeds from the catalogue that the seedsman claims to be disease resistant.

Varieties bred for resistance are increasing annually. It is also a fact that the rotation of crops referred to earlier is a major factor in reducing diseases.

Cleanliness

Finally, another factor worthy of mention and recommendation is the habit of cleanliness. The garden that is always 'spick and span' suffers less damage from disease than an untidy one. Always wash and clean your tools before they are put away. When a crop is harvested, all stems, leaves, etc should be cleared and put on the compost heap to rot. Make sure there are no weeds more than two weeks old!! Never allow diseased plants to remain in the ground with healthy neighbours.

When to plant and when to harvest

March is very much the beginning of the gardeners' year in most parts of Britain. It is the time when the hours of daylight begin to lengthen, and the sun gradually increases in strength and starts to warm the soil.

Nevertheless, in this early part of the year there are marked variations in the climate in different parts of the country, and a glance at our map will show how advanced or late the seasons can be. You will see that work that is done in the Midlands or the south of England in mid-March can be undertaken four weeks earlier in south Cornwall but not until three weeks later in the north of Scotland, and that there are other differences elsewhere. You will find that seedsmen and publishers issue charts showing when to sow, to transplant, to harvest, and so on.

Our own chart, we must make abundantly clear, is a broad guide to these activities, and each reader should take great care and relate its dates to the seasons as they fall in the part of the country where he lives.

Left: Cabbage 'Hidena'; tomato 'Pixie'; parsnip 'Hollow Crown'.

Gardening calendar as at mid-March showing approximate variants of climatic conditions in England, Wales and Scotland

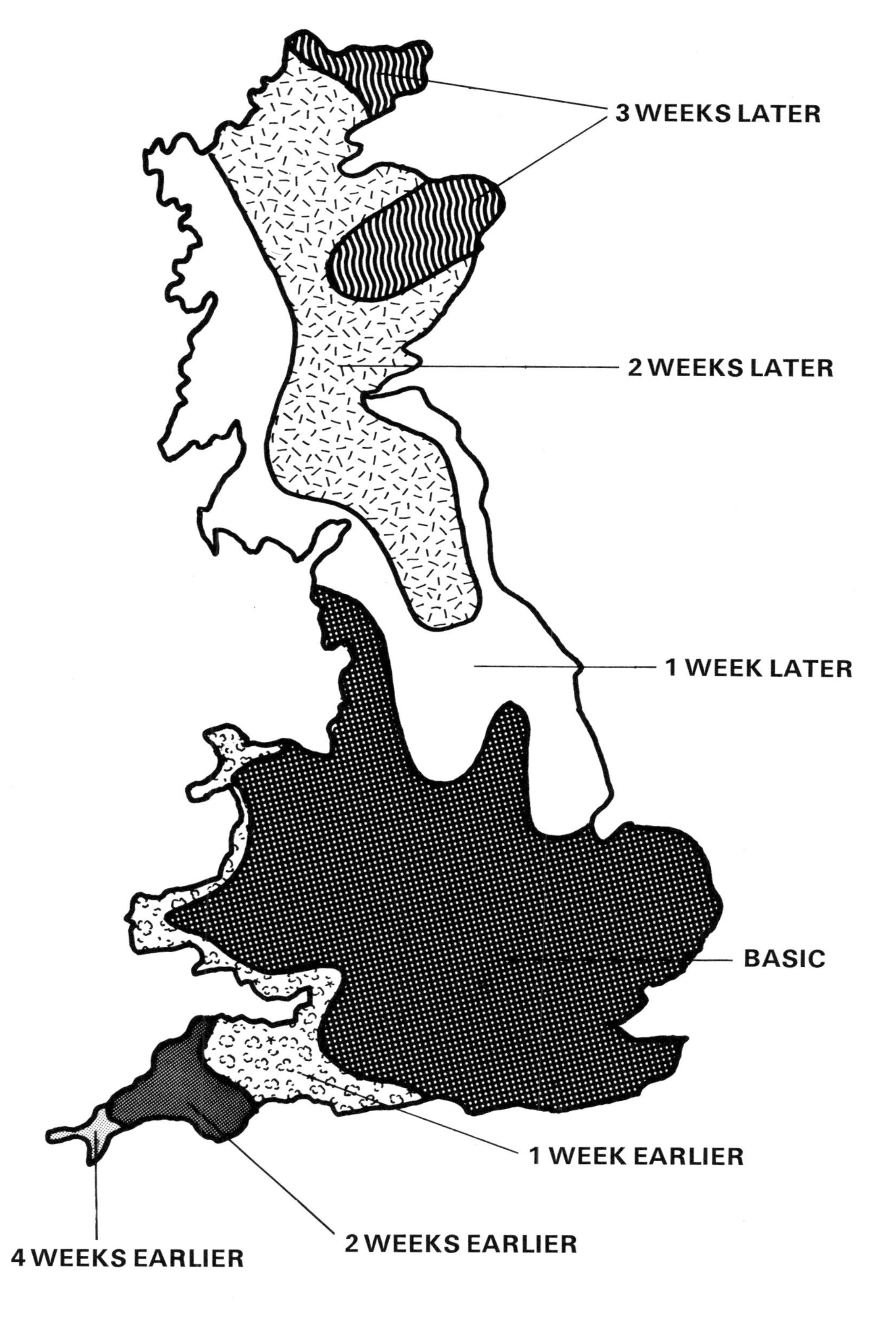

Vegetable growing calendar

All the popular vegetables are listed in the chart below, providing the information you need to help plan your vegetable garden. An 18-month period is covered so that crops becoming ready to harvest in the spring of the year following sowing can be shown. Please note that the times given are for average British climatic conditions, and will vary according to locality and weather conditions in any particular season. Plant spacings also depend to some extent on the varieties chosen. More detailed information is given on seed packets.

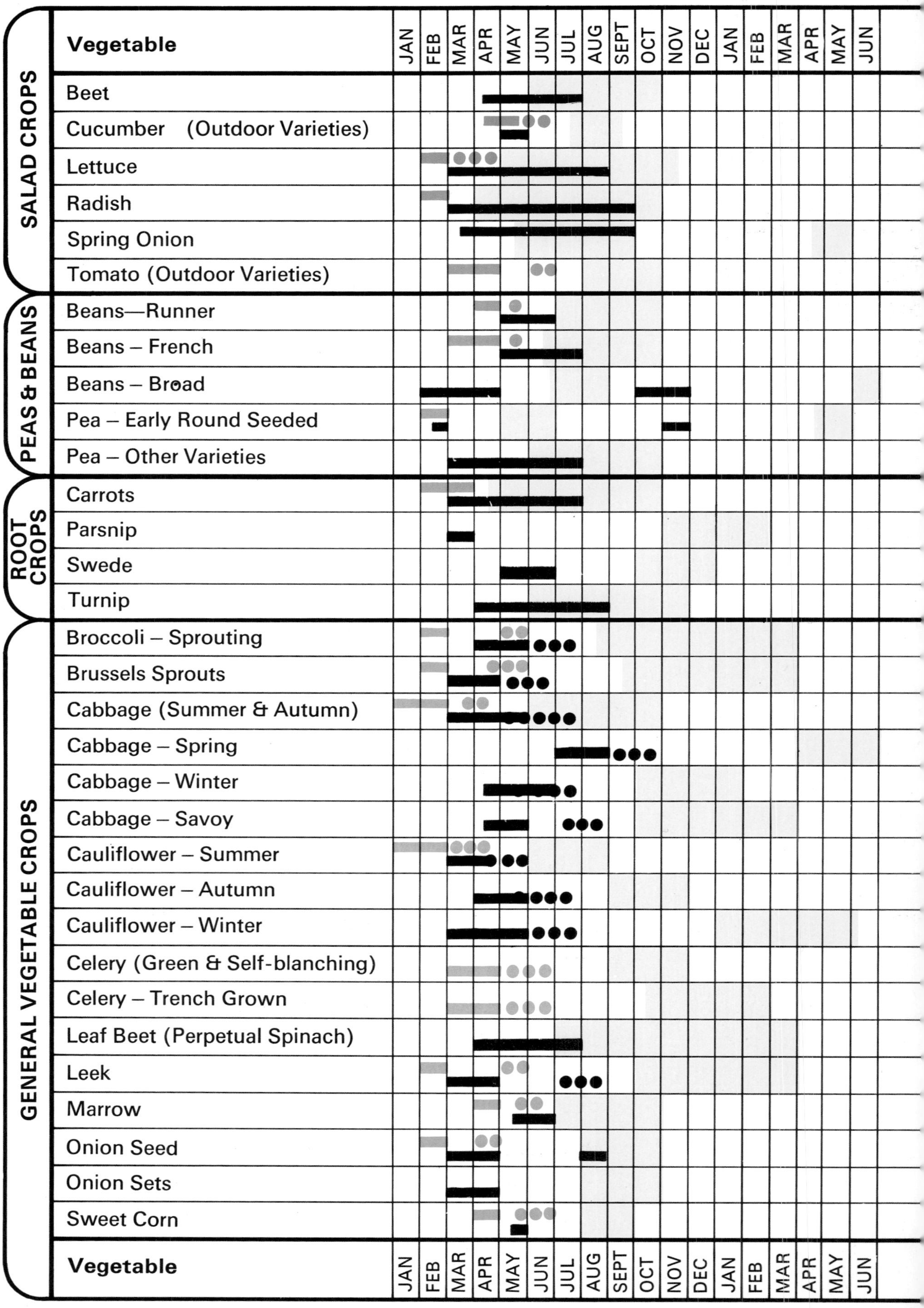

	Vegetable	JAN	FEB	MAR	APR	MAY	JUN	JUL	AUG	SEPT	OCT	NOV	DEC	JAN	FEB	MAR	APR	MAY	JUN
SALAD CROPS	Beet																		
	Cucumber (Outdoor Varieties)																		
	Lettuce																		
	Radish																		
	Spring Onion																		
	Tomato (Outdoor Varieties)																		
PEAS & BEANS	Beans—Runner																		
	Beans – French																		
	Beans – Broad																		
	Pea – Early Round Seeded																		
	Pea – Other Varieties																		
ROOT CROPS	Carrots																		
	Parsnip																		
	Swede																		
	Turnip																		
GENERAL VEGETABLE CROPS	Broccoli – Sprouting																		
	Brussels Sprouts																		
	Cabbage (Summer & Autumn)																		
	Cabbage – Spring																		
	Cabbage – Winter																		
	Cabbage – Savoy																		
	Cauliflower – Summer																		
	Cauliflower – Autumn																		
	Cauliflower – Winter																		
	Celery (Green & Self-blanching)																		
	Celery – Trench Grown																		
	Leaf Beet (Perpetual Spinach)																		
	Leek																		
	Marrow																		
	Onion Seed																		
	Onion Sets																		
	Sweet Corn																		
	Vegetable	JAN	FEB	MAR	APR	MAY	JUN	JUL	AUG	SEPT	OCT	NOV	DEC	JAN	FEB	MAR	APR	MAY	JUN

Sow under cloche, cold frame, or in unheated greenhouse

Plant out from under glass

Sow outdoors in open ground

Transplant outdoor sowings

Ready for use during this period

Distance between rows	Final Spacing in rows	Germination period (days)	Time to mature for harvest (weeks)	
15 in.	9 in.	10–20	16–18	Late sowing may run to seed
2½–3 ft	2½–3 ft	6	12–14	Stop runners when 3 in. (7.5 cm) long
9–12 in.	9–12 in.	6–12	9–10	Sow only winter hardy varieties in September
6 in.	—	6–7	7–9	Make frequent short-row sowings
1 ft	—	10–12	—	Pull as required
1½–2 ft	1½–2 ft	7–14	21 onwards	Keep staked
Double rows 5 ft apart	1 ft	7–14	16–18	8 ft. (2.45 m) stakes required ; pick regularly
1–2 ft	1 ft	7–14	12–14	Stake dwarfs with twigs, climbers with 5 ft (1.50 m) stakes ; pick regularly
Double rows 3 ft apart	9 in.	7–14	Spring sowing 14–16	Pick regularly
Same as Variety height	2–3 in.	—	25	Germination period will vary according to climatic conditions
	2–3 in.	7–14	14–16	
9–12 in.	4–6 in.	10–15	22–26	Use later thinnings in the kitchen
12–15 in.	6–8 in.	21–26	30–34	Lift and store any left in the ground in March
15–18 in.	9–12 in.	7–14	20–24	Protect from club root disease
15 in.	6–9 in.	7–14	11–13	
2 ft	2 ft	6–11	16–21	Pick regularly ; protect against root maggots
2½–3 ft	2½ ft	7–12	23–27	Protect against root maggots
2 ft	2 ft	7–12	16–18	Protect against root maggots
1½ ft	1 ft	7–12	16–18	Protect against root maggots
2 ft	2 ft	7–12	16–18	Protect against root maggots
2 ft	2 ft	7–12	16–18	Protect against root maggots
2 ft	2 ft	7–12	20	Protect against root maggots
2 ft	2 ft	7–12	16	Protect against root maggots
2–2½ ft	2–2½ ft	7–12	26	Protect against root maggots
9 in.	9 in.	9–21	20–24	
3–4 ft	1 ft	9–21	24–28	
1½ ft	15 in.	7–10	12–15	
1–2 ft	9–15 in.	6–12	26–30	Earth up gradually as they grow
4–6 ft	2 ft	7–14	12–14	
9–12 in.	6–9 in.	9–12	28–32	Sow exhibition varieties in January
1 ft	6–9 in.	—	25–28	
2 ft	2 ft	7–14	12–16	
Distance between rows	Final Spacing in rows	Germination period (days)	Time to mature for harvest (weeks)	

What type of soil do you have?

The following comments on each of the four main types of soil found in the United Kingdom, and the work that should be done in the spring, are printed by kind permission of Fisons Ltd.

Sandy soil
Light sandy soils are easy to work but they allow water to pass through too rapidly and, in doing so, much of the soluble plant food they contain is washed out. Peat will rapidly rectify these faults without in any way imparing the natural crumbliness of the soil. In the spring the addition of peat will improve fertility and there will be marked improvement in the use of fertilisers and water since both will be retained far better. Use peat 1 in. (2.5 cm) thick and manure at ½ lb (225 g) per square yard (square metre) and fork or rotovate both into the top 6 in. (15 cm) of soil.

Chalky soil
Soils containing much chalk or lime are too alkaline for some plants and are deficient in some essential chemicals which become locked up in insoluble forms. In winter they become sticky and difficult to work. Peat will correct all these faults and will also darken the colour of the soil, thereby improving its appearance and enabling it to absorb sun warmth more efficiently. Use Selected Garden Peat 1 in. (2.5 cm) thick at any time, and before planting add manure and fork or rotovate into the top 6 in. (15 cm) of soil to provide essential plant foods.

Clayey soil
Clay soils are usually potentially very fertile, but they are too cold and wet and are difficult to work. Peat will quickly open them up, making them easier to dig, so improving drainage, letting in air and stimulating healthy bacterial activity in the soil. Use peat in autumn or spring when the frosts have broken up the clods, 1 in. (2.5 cm) thick (twice this amount if the soil is very clayey) forked in the first instance into the top 4 in. (10 cm) of soil, but in subsequent years gradually increase this depth to 6 or 8 in. (15 or 20 cm). More peat can be given as a surface mulch in spring and peat can also be used in planting holes or scattered along seed drills. It is highly probable that this soil will require the addition of lime because laying wet it becomes acid.

Stony soil
Stones reduce the amount of useful soil and make it difficult to dig, fork and hoe. A compensating advantage is that, since stones remain relatively cool, moisture condenses on them and plants often cover them with a fine network of roots. So, though it may be necessary to remove all large stones from the surface, those below can be left, provided the fertility of the soil is maintained at a high level with peat and fertilisers. Any peat can be used according to the purposes for which it is required. Generous mulches of peat will cover stones, improve appearance and make it easier to hoe and weed. Manure applied in spring or summer will provide all necessary plant foods.

* * *

Double digging is usually considered to be more valuable than single digging, particularly on sandy, chalky and stony soils. It enables the addition of manure at a good depth where it will do most good for deep-rooted vegetables like brassicas.

This work should be done in the late autumn. Then, in spring, the soil should be broken down, with peat and manure added into the top 6 in. (15 cm), well raked or rotovated and made ready for planting out.

Clay soil, however, should be only single dug, but again have manure added. The trouble with double digging in this type of soil is that heavy clay is continually brought to the surface with the result, especially in damp weather, that working conditions are made difficult.

One of the best investments for a large vegetable garden is a rotovator which, when used in the spring and summer, breaks down the top 6 in. (15 cm) of soil into a smooth workable condition ready for immediate planting out. It will also mix soil, manure and peat into a good even mixture. Only stony soil presents any difficulty to the rotovator, and then only if the stones are very large. The advantage enjoyed by gardeners using a rotovator on heavy soil has to be seen to be believed.

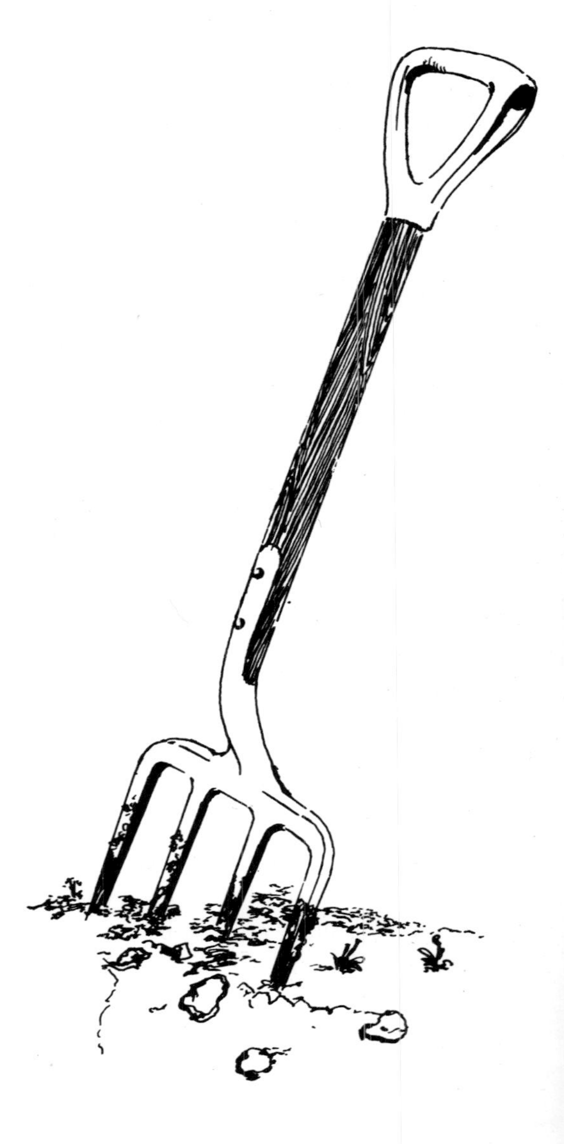

Sowing seeds and transplanting

It is desirable to arrange a small section of your garden as a seed nursery, which should be made ready as soon as the soil becomes workable – moist but not heavy. When this has been well prepared, with fertiliser added and well raked into the soil, firm down with your feet (avoiding heel marks). The soil should then be levelled off with a rake, but when raking do not make the soil as fine as dust otherwise it will make mud pies and, after watering, the seeds will be imprisoned in a tight crust of soil and unable to breathe.

Having prepared the nursery, it is essential to keep to the correct timing for seed sowing, but where the recommended time is, say, March or April, make two or three separate sowings to increase the period of harvesting. As to depth to which seed should be sown, follow the instructions on the seed packet (see also the chart on p 10 for spacing).

It is necessary to ensure that the seed bed is kept moist, but not overwatered. A good idea is to cover the seed bed with black polythene which will keep everything moist and assist with germination; but remove it when seedlings are above ground.

Another good suggestion is to 'scatter' sow. Make a trough of the correct depth for the particular seed 6 in. (15 cm) wide and scatter the seeds thinly over the area. This method avoids the necessity of early thinning and there is less chance of tangled, malformed roots as so often happens when seeds are sown in a thin single line. It is a method that is specially recommended for all very fine seeded vegetables.

Another method recommended is to sow three to six seeds in groups every few inches. Seeds do help each other up when soil crusts over them.

Weeds in a seed bed are a considerable problem, and all the work that is done to ensure that the seeds planted grow well acts equally to assist weeds in their growth. The only effective remedy is hand weeding as soon as the weeds have grown large enough to be distinguished from the vegetables. There is no short cut to overcome this problem and the earlier it is done, the less will be the disturbance to your seedlings.

Transplanting

When seedlings have grown sufficiently large to transplant into their permanent position in the vegetable garden, they should be moved with as little disturbance to the root as possible, keeping as much soil clinging to the root as can be managed.

Each hole into which the plants are transplanted should be well watered before planting and on the same day or next morning the whole bed should be watered to ensure that there is plenty of moisture and that the soil is around the stems and covering the roots naturally.

Readers who have cold frames, cold or heated greenhouses need not go to the trouble of having a nursery bed at all. They are advised to sow their seeds in pots or seed boxes and when seedlings are large enough to handle, prick them out into small pots of $2\frac{1}{2}$–3 in. These young plants, when large enough to be planted out, should be hardened off for a few days in the open before being removed from their pots and planted in their permanent positions. The great advantage of this method is that the seedlings are planted out without any disturbance whatever to their roots and, in consequence, suffer no setback.

When planting, dowse the roots and soil into a bucket of water before placing in the hole.

Above: Pea 'Senator'; 'Musselburg' leek; dwarf French beans 'Masterpiece' and 'Limelight'.

Crop rotation

The rotation of crops is a matter of sheer common sense and has been practised by farmers and horticulturists alike since time can remember.

The reasons are simple to explain. Various vegetables require ingredients of the soil in differing quantities. As an example, brassicas require large quantities of nitrogen and, if grown in the same soil annually, will not produce good crops after the first year. Beans and peas, however, feed nitrogen into the soil during growth. Hence, it is wise for peas and beans to be planted in the soil that is to be occupied by brassicas the following year.

Similarly, some plants attract pests and diseases which do not attack a different crop. If the same plot is used for the same vegetable year after year, the diseases and pests are there waiting.

Some plants grow better than others in soil that is heavily treated with well-rotted manure, but others prefer less manure and more general fertilisers, particularly shallow-rooted crops like peas.

Thus, rotating crops helps to keep the balance of plant food in the soil and reduces the danger of pest and disease problems.

Above: Broad beans.
Right: Pea 'Show Perfection'. SUTTONS SEEDS LIMITED
Peas and beans are crops that are well followed by brassicas.

Chart for Crop Rotation in a Vegetable Garden

Year 1

OTHER 1 CROPS	ROOT 2 CROPS	BRASSICA 3 CROPS

Year 2

BRASSICA 1 CROPS	OTHER 2 CROPS	ROOT 3 CROPS

Year 3

ROOT 1 CROPS	BRASSICA 2 CROPS	OTHER 3 CROPS

Helpful hints

★ Many people do not have heated greenhouses and those who do find it increasingly expensive to keep them heated above 45°F (7°C). The boiler house, or one of those warm places that we all have indoors, can be put to good use for propagating seeds.

★ Plastic cups from the office or works coffee machine, which are usually thrown away after use, make excellent pots for pricking out. Many other containers that can be used, like yogurt or cream pots, can be saved in the house.

Boxes for seeds and pricking out are expensive, but shopkeepers, especially fishmongers, are usually pleased to sell their old boxes for something less than the cost of the regular thing.

★ Plan the vegetable garden so that rows run right across from north to south, but plant, say, only a third of a row of any one vegetable at a time. Planting at intervals will help to give a continuous supply, which will help to reduce the waste that can occur when all of a crop ripens at the same time.

★ Protection against birds can be very expensive. To make 'scarecrows' by putting sticks in the ground and covering the tops with sheets of paper that will blow about can be effective. Better still is to make some hurdles, say 4 ft (1.20 m) high and drape nylon over them. The hurdles, being movable, can be shifted from row to row, and the nylon, if treated with care, is almost everlasting. Even so, it is wise to replenish the stock every year.

★ Money spent on slug bait might seem to be an extravagance, but it saves disappointment and loss, and is, therefore, a 'must'.

★ Even vegetables from the garden will lose some of their flavour when they have been picked for several days. If you have a surplus you will find that your neighbours or colleagues will be delighted to buy from you – especially if you charge a little less than the current shop prices. The money, of course, goes towards the cost of seeds, netting, boxes and so on.

★ You may not have a cold frame; but you do have a small corner of the garden that is unsuitable for cultivation? Why not build a cold frame yourself? All you need (except for the top) are a hundred or two bricks and a couple of bags of cement. If a sliding top, or even one that rises on hinges and can be propped up at varying heights, is too difficult, a sheet of heavy polythene may be used as a cover. Suitable dimensions are: back wall 3 ft (90 cm) high, front wall 1 ft (30 cm) high, and the width whatever can be spared from the odd corner of the garden. It is best for the tall back to be to the north so that the sun can shine right into the frame.

★ Remember to rotate your crops (see page 14).

★ Do not throw away old newspapers, rags, dusters and other waste materials. Such items are invaluable for retaining moisture in the soil, particularly under rose trees, at the base of the trench for runner beans, or when 'dug in' during autumn digging. Newspapers and rags should be torn into small pieces and well dampened in a bucket of water. If possible, add a little salt of potassium, magnesium or iron; but even without, the result will be a greatly improved crop.

★ Do not let watering the garden become an unwelcome chore, for nothing is easier. Twelve hours of watering need not take up more than ten minutes of your own time in a day. You can buy a specially-made hose that lies flat on the ground and has dozens of small holes through which gentle sprays of water cascade. The holes are cut in such a way as to send the water in several directions to cover a large area rather like a good light shower of rain. After setting up the hose, all you have to do is to move it occasionally. Do not fall into the trap of allowing plants to be overwatered, especially in very heavy soil.

You could make such a hosespray yourself by drilling holes in an old hose, but it is not likely to be as effective as a purpose-made device, and the stopping of one end could be a problem.

★ How to measure a bushel. A bushel is the normal measure used in gardening, particularly when mixing soils for potting and seed sowing, but few manufacturers sell their products by the bushel. A simple way to work it out is to fill a bucket with two gallons of water and mark the water line – then fill the bucket four times to this line, and you have a bushel. Remember: a bushel is quantity and not weight. For example, a bushel of peat weighs very little, but a bushel of grit is very heavy.

★ Make sure that you have the latest catalogues from the leading seedsmen. They will keep you up to date, give you fresh ideas and provide valuable items of knowledge that might not otherwise have come your way.

Right: Lettuce 'Susan'. SUTTONS SEEDS LIMITED

Should I have a greenhouse?

This is a very important question and you will want all of its aspects to be answered to your satisfaction before you embark upon such a project, because a greenhouse

★ is, in most cases, costly to buy,
★ is expensive to heat,
★ takes up a lot of time,
★ requires maintenance.

If you are at all interested in gardening a greenhouse is one of the finest investments you can make, not just for the financial return you obtain, but also for the enormous interest it engenders and the great satisfaction it gives. Some of the advantages that a greenhouse brings are summed up by saying:

★ You can grow a large quantity of fruit, such as tomatoes, grapes, melons and so on.
★ You can keep your home fully supplied with flowers all the year round.
★ You can extend the growing season of vegetables and bedding plants through being able to propagate crops earlier.
★ The wet and cold days of winter need not be dreaded or lost because there is always work to be done in the greenhouse – where it is both dry and warm.
★ Whatever the original outlay on a greenhouse, provided you are prepared to give your time freely and plan sensibly you should soon be able to save more on housekeeping expenditure alone than the original cost.
★ For people in retirement it is one of the greatest contributors to a longer and happier life. It ensures that there is always something to do and, more important, something to look forward to – the first tomato to ripen, the first chrysanthemum to bloom.

If you feel, by now, that the advantages far outweigh the disadvantages and decide to go ahead with a greenhouse, the following points may be helpful.

Positioning is important. The greenhouse should be placed where it will enjoy the maximum light, and preferably with the sides facing north and south and the ends east and west, so that it gets the greatest amount of sunshine. It should not be placed too near tall hedges or trees that could obstruct the light – trees in particular can shade the sun for part of the day, especially in the spring, autumn and winter when the sun is low in the sky. For preference, the door should face west away from cold east winds.

It is advisable to decide before buying the greenhouse just what types of plants, fruits and vegetables you intend to grow.

You may wish to specialise in two main crops – perhaps tomatoes in spring and summer, and chrysanthemums in summer and winter – with only a few pot plants and vegetables. In that case you will require good soil for the tomatoes and chrysanthemums, and a limited amount of staging for the pot plants and seed boxes. On the other hand, you might want to have lots of staging and only a small space for tomatoes and chrysanthemums. In any case, you should plan how you intend to use the greenhouse and then have a good look round *before* deciding which greenhouse to buy.

Heating is also a problem to be considered. What temperatures are you going to need? Will you want lighting for evening work?

To summarise, the following questions require answers:

★ What size and shape of greenhouse do you need?
★ What material should the greenhouse be made from – timber or aluminium and glass, or polythene?
★ Does the site selected provide the maximum amount of sunshine and daylight?
★ What heating is required, and by what method – gas, electricity, solid fuel or paraffin? If gas or electricity, can it be supplied from the house? Will the greenhouse be near enough to the house to 'tap' your own central-heating system?
★ What plants are to be grown? What height will they require? (tomatoes and chrysanthemums particularly). How much staging will you need?
★ How often are you away from home? Will you need some form of automation for vents, heating and watering?

Concerning timber or aluminium, there is very little to choose between them, but aluminium does not need as much maintenance as wood and is easier and lighter to erect, but possibly costs more to heat in the winter.

If cost is of considerable importance, you might care to consider a plastic greenhouse. Experience shows that, provided the polythene is of good quality, these greenhouses can be cheap and efficient. The fact must be borne in mind, however, that the polythene will have to be replaced every two years.

Another problem is that plastic does cause considerable condensation, but for many things that should be grown in a greenhouse that does not present any great difficulty.

A reasonably competent 'do-it-yourself' man could quite easily build himself a greenhouse using a structure of either rigid tubular steel or seasoned wood (which might be obtained from a demolition merchant) and polythene. He should remember to avoid sharp edges that could slice the polythene.

Simple forms of walk-in polythene tunnels are also available on the market. They can be immensely useful, for they are light in weight and portable, and can be moved around the garden and placed just where they are wanted to protect or force a crop. As they vary in size, the full range should be considered before a purchase is made.

Finally – whatever size you settle for, whatever type of greenhouse you buy, you will always wish that it was a little larger; so always incline towards larger rather than smaller size. Some greenhouses are made so that another section can be added to the original structure at the end opposite the door. In this way you can increase the size of your greenhouse without too heavy a cost at any one time. Bear this in mind when you plan your first greenhouse and leave room on the site for expansion.

How to garden for profit

Much will depend upon your enthusiasm, the time you have available and the size of your garden, but there is no doubt that every hour you spend in the vegetable garden will put money in your pocket.

The prices of vegetables have been soaring upwards and will continue to do so. Why? There are a number of reasons:

★ The cost of fuel to heat growers' greenhouses has risen enormously.
★ There has been a very large rise in horticultural workers' wages.
★ Fertilisers, insecticides, seeds and compost all cost more.
★ Glass and repairs to greenhouses are considerably more expensive.
★ Growers, too, have to face the increased cost of living and require a higher return on their capital to maintain their living standards.
★ Many growers have gone and are going out of business, chiefly for reasons connected with the financial problems listed above. This has caused shortages and a consequent rise in prices. The trend is not likely to change.
★ Growers no longer benefit from a government subsidy.

What can be done to avoid the continuing increases and stop the weekly housekeeping bills getting out of hand? The answer lies in your vegetable garden and your own efforts. Moreover, there will be a bonus in the excellent flavour that is to be found in home-grown vegetables.

There can, however, be more to the vegetable garden than simple economising. Inevitably there will be a surplus in certain crops, and these surpluses can be put to making a profit. The vegetables that are likely to provide a surplus are listed below. Look at the cost of these vegetables in the greengrocer's shop and you will see immediately how valuable the sale of surpluses from your own crops could be to you.

Your outlet could be somebody who lives in a flat, or a colleague at work who does not have a garden, and would welcome some fresh vegetables. In the summer you could easily add £3 or £4, or even more, to your weekly income – quite apart from the savings in the shopping bills from growing your own vegetables.

In the world as a whole there is an increasing shortage of food, and the days of imported cheap food are over. Truly, 'every hour you spend in the vegetable garden will put money in your pocket'.

Some crops that will give a surplus

Crop	Season
Broad beans	Summer
Brussels sprouts	Autumn, winter
Cabbage	Spring, summer, autumn, winter
Carrots	Summer, autumn
Cauliflowers	Summer, autumn, winter
French beans	Summer, autumn
Garden peas	Summer
Lettuce	Spring, summer, autumn
Peas	May to September
Runner beans	Summer, autumn

Vegetables

On the following pages fairly extensive details are given about individual vegetables and how to obtain a good crop. But before that there are a number of points to be made because no one collection of advice can possibly be completely apt for everybody – if only because of the different conditions that are to be found in various parts of the country. Then, locally, soils vary enormously; some gardens are open to the elements, others are protected; some face north, others south, and so on. These factors should be borne in mind when reading the advice that follows.

Lime

Many parts of Britain have a heavy or clayey type of soil, and it is there particularly that the addition of lime is required. In chalky areas lime is seldom, if ever, necessary.

Lime should never be added at the same time as manure or fertilisers. It can be added in the autumn provided no animal manures or chemical fertilisers are being added at the same time.

The rule is that lime should, where necessary, be added from six to eight weeks after manure has been dug in, and fourteen days, at least, before fertilisers are distributed in the spring. This is because lime releases the ammonia from manure and that is required to remain in the soil for vegetable growth. Lime also reacts unfavourably on some chemicals. Do not use too much lime – 1 oz (25 g) to a square yard/metre is about right.

Manure

If a garden is being developed from virgin soil the soil must be well dug and broken down before manure is added. If that is not done the two will never mix: the manure will remain in isolated pockets and so will heavy type soil, with the result that plants will have only the choice of feeding on manure or clay – both of which they dislike on their own.

Horse manure is best for lightening and opening up heavy soils.

Cow manure is heavier and wetter and therefore preferable for light and sandy soils.

Pig manure should be used only on the very lightest of soils, and then only when mixed with plenty of organic refuse all of which must be completely decomposed.

A mixture of all three manures is excellent because between them there is the balanced food that most vegetables need.

The droppings of birds, sheep, goats, rabbits and chickens should be used as fertilisers rather than as manures.

Hop manures are usually proprietary products in an extremely concentrated form. They must not be used in nearly such large quantities as ordinary manures.

Moss peat is intensely valuable for breaking up heavy soils but, of itself, does not add to the goodness in the soil. Most proprietary peats have fertilisers added for that purpose.

Seaweed has a high reputation as a manure and contains a large percentage of potash. It is, therefore, particularly beneficial in very light soils. Cabbages and beetroot do especially well where seaweed is added to the soil.

Liquid manures. Because all plants derive nourishment from the soil in a solution liquid manures are frequently recommended. They supply nutrients in a form that can be immediately absorbed through the roots of both young and established plants.

The compost heap

The materials that make the best compost heaps include sods; grass cuttings; green leaves; pea, bean and potato tops; leaves of vegetables such as cabbages; some hedge clippings, provided that the stalks are not thick and woody; droppings from chickens and other birds; and especially the ashes from a bonfire. Do not use cinders from a coal fire, sawdust or oily substances.

To keep a good, neat compost heap it is advisable to drive four stout stakes into the ground and run 3 ft (90 cm) of wire mesh round. This stops the compost spreading and so helps in the process of rotting.

Organic and inorganic fertilisers

Organic fertilisers are those which are derived from animal or vegetable matter. Inorganic fertilisers are chemicals and are sold as proprietary products. They are valuable additions to the soil, but in no way are they substitutes for organic fertilisers which not only feed plants but help to improve the soil and retention of liquids. Inorganic fertilisers should only be used in a supplementary manner, and then with care – never exceed the dosage recommended by the manufacturer.

Agricultural salt

This is excellent on all soils. On light soil it helps to hold moisture, and on heavy land it assists in releasing potash. A light dressing of 1 oz (25 g) per square yard/metre over the whole vegetable garden will have a very beneficial effect.

Chemicals

It is very unwise for an amateur gardener to use separate chemicals unless he really knows what he is doing – and why. Proprietary fertilisers, which have ingredients expertly mixed in the correct proportions, are much safer and usually more effective.

Applying fertilisers

Always rake fertilisers into the soil to prevent them clogging into lumps on the surface. If the weather is dry, either dissolve the fertiliser in water before application or apply in the manufactured state and water afterwards.

Rotovators

The value of a rotovator has been mentioned elsewhere in this book, but it is worth repeating that anybody planning or owning a reasonably sized garden will find the purchase of a rotovator a wise investment. It will reduce the time spent on work by many, many hours and take a great deal of backache out of the process of preparing the ground.

Above: Artichoke 'Grande Beurre'.
THOMPSON AND MORGAN
Right: Globe artichoke.

Gardens that face north and east

Crops in such gardens need some protection and it is advisable to erect a good fence or grow a strong hedge to prevent ill effects from cold winds.

Where you live

Study the map on page 9 which gives a rough guide to the differences in the country of the climate in mid-March, or early spring. Then consider that the seasons vary even from year to year. A typical English summer may be hot, cold, wet and dry in quick succession, and sometimes much warmer in the north than in the south. Indeed, the gardener's greatest problem is the fickleness of the English climate. But there is little consolation in the fact that sometimes crop failure is due to circumstances far beyond human control.

The choice of varieties of vegetables

Generally speaking, in the pages that follow no advice is given on particular varieties of any one vegetable, for the requirements of individual gardeners vary enormously. Seedsmen, however, give full descriptions of each variety that they list in their catalogues, and growers can easily make selections to meet their own needs.

The different crops

Root crops	*Brassicas*	*Shallow-rooted & other crops*
Beetroot	Brussels sprouts	Broad bean
Carrots	All cabbages	Celery
Chicory	Cauliflowers	Leeks
Jerusalem artichokes	Kale	Lettuce
Kohl rabi	Savoy	Onions
Parsnips	Sprouting broccoli	Peas
Potatoes	*Trench*	Spinach
Radishes	French beans	Sweet corn
Swedes	Runner beans	Tomatoes
Turnips		

Rooted crops and brassicas need plenty of manure at 10 in. (25.5 cm) depth, but carrots and brassicas (especially carrots) should be planted where the manure was dug in a year earlier.

French and runner beans should be planted in a trench 4 in. (10 cm) lower than normal soil level for ease of watering during growth. When the trench is dug 12 in. (30 cm) deep, fill in with plenty of manure, old newspapers, grass cuttings and peat, and cover with 3 in. (7.5 cm) of good top soil. Once the roots find their way down to the compost, the plants will thrive.

Shallow-rooted crops require some dry manure and fertilisers raked into the top soil to produce really good crops.

Globe artichoke

The globe artichoke is grown as a summer vegetable for its edible head (which has large, fleshy scales) unlike the Jerusalem artichoke which is grown for its tubers. The entire head should be cut when in the bud stage, from the end of June.

After boiling it may be served hot or cold. The edible part is found at the base of the scales and the heart has, perhaps, the best flavour.

It is a perennial plant re-appearing year after year, but it needs protection from frost.

How to grow: The plants should be placed in a sunny position and in good, fertile, well-drained soil which has been well manured. More manure should be added each late autumn.

The crop may be increased by severing rooted suckers in November. The suckers should be potted in potting compost, kept away from frost and planted out at the end of April at intervals of 3 ft (90 cm).

Although the globe artichoke is a perennial it is advisable, to ensure good crops, to discard plants after four years and renew with the new-rooted suckers.

Pests: The plant is virtually free from trouble.

Diseases: *Petal blight.*

Jerusalem artichoke

This is not a very commonly used vegetable but it is easily grown. It requires no special feeding or manuring and will grow in most well-drained soils.

The Jerusalem artichoke is grown for the tubers that it produces. They are of high quality and have an extremely good flavour when eaten young. The plant will grow to anything between 6 ft and 12 ft (1.80–3.65 m) high.

How to grow: Plant the tubers (similar to potatoes) in February or early March, setting them 1 ft (30 cm) apart and 6 in. (15 cm) deep. The tall foliage, which grows from 7 to 12 ft (2.10 m–3.65 m) high, should be pinched back in summer to prevent flowers forming. Tubers may be left on the plant, but they can be lifted in the early winter and stored in sand.

Cooking times vary according to size, but the vegetable should be boiled or steamed before peeling and peeled only when ready to serve.

Pests: *Slugs* and *caterpillars.*

Diseases: *White flea fly fungus.*

Asparagus

Asparagus is looked upon as one of the luxury vegetables – delightful to eat and very expensive to buy. From the gardener's point of view, the first thing about asparagus is that its bed is a permanent part of the garden. Many asparagus beds now in existence have been active for more than a hundred years. A new bed, if made properly, should outlive the older members of the household, for it should last for at least thirty years.

The next point to consider is that an asparagus bed is not really appropriate to a small garden. It takes up a lot of room and prefers to be exclusive, although in the later season it may be used to grow catch crops of lettuce or carrots between the rows.

To make an asparagus bed can be a formidable task. Great care is required to make a good one – which is the only thing to do if it is to last a lifetime.

Soil: The natural for asparagus is sandy. Most of the famous growing areas, such as Formby Sands, have sandy soil, but asparagus will grow in most types of soil given a little encouragement.

Effective drainage is the first essential; and after that double trench digging, preferably with a 6–12 in. (15–30 cm) layer of pebbles beneath.

The bed should be given a very open position where it will receive the maximum sunshine. If the soil is very heavy – such as clay – it should be removed altogether and replaced with a mixture of loam soil, leaf mould, some manure, sand, some gritty material and the addition of bone meal and a fair dusting of lime. When finished, the bed should be in the shape of a dome, the centre being at least 12 in. (30 cm) higher than the edges, and either 3 ft (90 cm) or 4 ft 6 in. (1.35 m) wide. There should be a gangway 2 ft (60 cm) wide between beds.

Another way to make a bed, if the soil is excessively heavy, is to build a raised bed. There is much to be said for planting in soil above ground level, which can consist of special mixes and have excellent drainage. Having selected the site, build a frame made of timber, brick or some other suitable material around the area about 1 ft (30 cm) high and fill it with the ideal soil described above. This method is very popular in the United States where it is much used to make beds for many varieties of vegetables and for herb gardens that are conveniently placed near the kitchen.

How to grow: In the narrower bed allow two rows of asparagus, and in the wider bed three rows. Only when the beds are prepared, and NOT before, should you decide how to fill them. There are two methods, rather like Victor Sylvester's 'Quick, Quick, Slow'. The quick method (it is not the less expensive, but it is not excessively expensive) is to buy plants that are one, two or three years old. Three-year-old plants bought one year will mature the next year, and two-year-old plants need two years to bear a crop. The slow method, which is the cheaper method, is to sow seeds, but worthwhile harvesting must not be expected for four years.

Planting and sowing: If seeds are used they should be sown in a nursery bed in April and transplanted into the asparagus bed a year later.

The planting season for one- two- or three-year old plants (or your own seedlings) is late March or April according to the part of the country. It is most important that roots should not be exposed to the air for one minute longer than is necessary. For this reason, the bed *must be ready* when roots are bought, and as the packages are opened planting must follow immediately. Provided the rules are followed, planting is a simple matter.

If the bed is 3 ft (90 cm) wide, take out drills 9 in. (22.5 cm) from the long sides; for the 4 ft 6 in. (1.35 m) wide bed add a drill down the middle. As the roots resemble the legs of a spider – but are proportionately longer – it is necessary when the drill is taken out to build a ridge down the centre of the drill for the plant to sit on and allow the roots to trail down the sides like somebody astride a horse. The ridge should be deep enough to allow the crowns of the plants that are in position to be covered with 5 in. (12.5 cm) of soil. Cover each root as it is planted.

The first year's entire crop of 3-year-old roots (longer for younger roots) must be allowed to grow without cutting and be left to become the well-known asparagus fern. When this fern turns yellow in the late autumn, cut it down to ground level, rake the bed lightly, clean out all weeds and dress with about 3 in. (7.5 cm) of manure all over, leaving this dressing on top. Remove the manure in early May and give the bed a dressing of agricultural salt, 4 oz (100 g) per square yard/metre.

Harvesting: To cut asparagus, use a sharp, pointed knife. Thrust the knife downwards through the soil and cut about 2 in. (5 cm) below the surface and close to the shoot that is to be cut. Take great care not to harm the crown or young shoots below the soil level. Harvesting should cease at the end of June or very early in July.

Forcing asparagus: Early crops can be obtained by planting established roots in a heated frame. The soil should be well prepared in advance. The roots should be lifted from the asparagus bed in October and immediately replanted in the heated frame in a 4 in. (10 cm) bed of leaf mould, and with sifted soil worked around the roots and crowns and just covering the crowns.

Two weeks later cover the whole bed with 6 in. (15 cm) of similar soil and cover its surface with a layer of peat. The first shoots should appear within a month and cropping should continue until after Christmas. The temperature should be 60–65°F

Right: Asparagus 'Royal Select'.

(15.5–18.5°C), and the frame should be kept closed. After cropping, the roots used for this purpose should be destroyed as they will be of no further use.

Pests: *Asparagus beetles* feed on young shoots and foliage, and make plants sticky and black.

Disease: *Violet root rot* causes top growth to turn yellow and die long before the autumn.

Frost damage can be caused if the top dressing of manure is removed too early.

Broad beans

Seeds: The size of garden, whether it is open to winds or protected will have an important bearing on the type of seeds that should be sown. There are three main varieties from which to choose:

The Long Pod varieties which are recognised by their long narrow pods and are generally regarded as the hardiest, earliest and heaviest croppers. Suitable for sowing in the autumn.

The Windsor varieties have shorter, broader pods and are regarded as the best for flavour. They mature later than the long pod variety. Suitable for Spring sowing.

The dwarf varieties have short pods and a bushy growth. They are regarded as the best variety for small gardens, particularly those exposed to heavy winds.

Sowing: Seeds sown in the autumn give a good early crop which is seldom attacked by aphids (black fly) which are so prevalent among broad beans. The time between sowing and harvesting of this autumn sowing is approximately twenty-six weeks, whereas the spring sowing matures in approximately fifteen weeks. The spring sowing can commence in late February and additional sowings can be made up to as late as the end of May.

It is wise not to sow all the seeds at the same time, and intervals between sowings will extend the harvesting period and thus avoid waste by having too many beans at any one time.

When sowing, make a shallow trench 12 in. (30 cm) wide and 2 in. (5 cm) deep and place a row of seeds each side of the trench about 6 in. (15 cm) apart. Replace the soil and firm down with the back of the rake.

Leave at least 2 ft (60 cm) between this and any other crops to allow for the broad beans' growth and room for the person harvesting.

Soil: Broad beans grow best on rich and heavier types of soil, which have been well fed with manure for some previous crop, though this need not have been dug in as deep as for brassica crops because the beans are by no means so deep rooted.

However, adequate crops can be grown on almost any soil provided it is neither waterlogged nor acid. Application of lime will prevent the latter and a good digging in the autumn should prevent waterlogging. When the ground has been prepared ready for sowing, an application of a good general fertiliser should be made one or two weeks before sowing takes place.

Crop maintenance: The crops should be kept moist and well watered in dry weather. Tall varieties should be supported with sticks at 8 ft (2.45 m) intervals with one or two continuous lines of string running round rows. When the plants are fully grown prick out the top 3 in. (7.5 cm), which will help mature the crop and prevent attacks by aphids. (If these tops are not affected by aphids they may be cooked as a pleasant vegetable. They retain the flavour of broad beans.)

If crops are affected by aphids they should be sprayed with the appropriate insecticide.

Harvesting: For the most flavour the beans should be picked when young and they should on no account be allowed to grow to their full size, except for the few that are being left for next year's seed. If, when podding, it is seen that the scar on the bean itself has turned black, they have been left too long. This scar should be white or green at the time of picking.

Harvesting time from the autumn sowing begins in early June and spring sowings will extend the season to as late as early September, according to when sowings are made.

Pests: Seeds may be attacked by *slugs, millipedes* and *bean seed fly maggots. Aphids*, particularly *black fly*, infect the terminal shoots of young and established plants. *Capsid bugs* feed on young tissues, causing tattered leaves and blindness of buds. In dry weather, *glasshouse red spider mites* may infest leaves, which change colour and wither.

Diseases: *Anthracnose of dwarf bean* causes brown spots on leaves and stems and brown sunken areas on the pods. *Foot and root rot* is caused by various fungi. Roots die and show black patches; the stem base becomes discoloured and dedecays, foliage turns yellow. *Fusarium wilt* causes leaves to turn yellow and plants to wilt. The stem bases and roots show red marks. *Grey mould* attacks the pods in waterlogged soil and very wet weather.

Lack of moisture and cold nights prevent flowers from setting.

Above: Dwarf bean 'The Prince'.

Left: Broad beans.

Deep freezing: Shell beans and wash well. Blanch for 3 minutes, cool and pack in polythene bags.

French beans

A great favourite with everybody who enjoys vegetables. By far the greatest number of gardeners grow the dwarf green varieties, which grow only to 12 in. (30 cm) high and do not require sticking for support. More and more people, however, are beginning to grow the climbing variety which do need supports and grow to a height of 5 ft (1.50 m). The swing to the climbing variety is probably because the dwarf variety grow so close to the ground that the beans are inclined to be on the soil and, after heavy rain, become covered in soil and, in some cases, damaged by insects. Because they grow so low down, it is a backaching job to pick them and they are sometimes difficult to see.

The climbing variety are equally prolific in cropping and of excellent flavour, but should be given a sheltered position away from winds.

These two varieties are the most popular groups and are available in either flat or pencil podded varieties. The former variety, if left a day too long before picking, can become hard and stringy, the pencil podded variety are guaranteed stringless.

In addition to the above there are the yellow and purple varieties. The great advantage these have over the green varieties is at harvesting time, when they are much more easily seen, contrasting as they do with the leaves and storks of the plant.

Seeds: Study the seedsman's catalogue with care before ordering and decide which variety you require, green dwarf, green climbing, yellow or purple. Much will depend on the space available when deciding between dwarf and climbing varieties – clearly the dwarf will require the smaller area, which is largely the reason for their popularity.

Soil: To obtain the best results, the soil should be dug at least 10 in. (25.5 cm) and given a good dosing of well-rotted manure or compost (1½ buckets per square yard/metre) and a light dressing of lime in October or November. Two weeks before sowing rake the ground thoroughly adding 2 oz (50 g) per square yard/metre of a general fertiliser.

In the case of climbing varieties, the very best results will be obtained if a trench is dug 12 in. (30 cm) deep and treated as above, but with the surface of the trench left 4 in. (10 cm) below normal soil level to assist with watering during the growing period. It will also assist rain to drain down into the roots.

Seed sowing: Seeds should be sown 2 in. (5 cm) deep into the top soil in late April or early May, but do not sow all the seeds at the same time, thus extending the period of cropping. If a warm greenhouse is available, sow the early part of the crop in individual pots, which when 4 in. (10 cm) high can be transplanted into the bed without any set back, after having been hardened off in the open. They are susceptible to frosts so this must not be done until the likelihood of frosts is passed.

Care of the crop: Support dwarf varieties with small twigs and climbing varieties with 5 ft (1.50 m) twiggy branches or plastic netting stretched along the row. Spray regularly, once flowers have formed, with a fine spray of water to assist setting – watch for black fly on blooms and if seen spray with insecticide immediately or they will spread over the whole crop. Use slug bait or pellets against slugs which will attack seedlings. Hoe and water regularly.

Harvesting: Commence picking approximately twelve weeks after sowing when pods are about 4–5 in. (10–12.5 cm) long. The test of readiness is when pods snap easily, but before seeds begin to form up into bulges. Regular picking several times a week will ensure a continuous crop for six to eight weeks.

Haricot beans can be obtained by leaving the last batch of beans to grow on the plant until they turn a browny yellow. Hang up the plants to dry in a dry place. When the pods are completely dry and brittle, shell the beans and finally dry them on sheets of paper for several days. The beans should then be kept in an airtight container.

Pests: *Slugs, millepedes* and the *maggots of bean seed fly,* eat seeds or seedlings. *Aphids,* particularly black fly, infest terminal shoots of both young and established plants. *Capsid bugs* feed on young tissues, causing damaged leaves and bud blindness. *Glasshouse red spider* may infest leaves during excessively dry weather.

Diseases: Brown spots on leaves or stems are usually caused by *anthracnose of dwarf bean* which will go on to cause brown sunken areas on the beans. *Foot and root rot* is caused by fungi and roots will die, stem bases becoming discoloured. *Fusarium wilt* causes leaves to turn yellow and the plants to wilt. Stem bases show red marks. *Grey mould* attacks pods in very hot weather. *Frosts* will kill and *cold winds* will prevent blooms setting.

Deep freezing: French beans are excellent for deep freezing. Wash well, top and tail, blanch for three minutes. Cool and pack in polythene bags.

Runner beans

A highly enjoyable vegetable and most gardeners regret that its harvesting time runs parallel with that of the French bean, but both are suitable for deep freezing, though perhaps the French bean is the better of the two after freezing.

The majority of runner beans, or 'scarlet runners' as they are generally called because of their attractive red flowers, grow to a height of 8–10 ft (2.45–3.00 m)

Right: French bean 'Provider'.

and bear beans which are from 10 in. (25.5 cm) to 20 in. (51 cm) long, according to the variety chosen. They have to be supported on good strong stakes not less than 8 ft (2.45 m) long which should cross one another approximately 4 ft (1.20 m) from the ground and be bound tight to another stake running at right angles through the middle where the upright stakes cross. Good strong string should be used to hold these stakes in position because the weight of foliage and beans that grow above the level of 4 ft (1.20 m) is considerable and, unless the supports are well made, winds could blow the whole construction down. It is for this reason that the most protected position in the garden should be reserved for runner beans, coupled with the fact that the beans themselves, if blown about, bruise easily and turn black where bruised.

A few varieties can be grown as short bushy plants, although these too are normally tall growing, but by pinching out the main stem when plants are about 12 in. (30 cm) high, bushiness is induced. The side shoots should, as they grow, be pinched out every seven days, and the stems supported with short twigs.

The resulting pods are by no means as perfect as those grown normally as this method can only be recommended to those who live in particularly windy regions, such as the north coast of Cornwall. For those subject to such strong winds the dwarf runner bean is probably preferable. Plants grow naturally to 18 in. (45 cm) high and produce pods 8 in. (20 cm) long. Short support sticks are desirable.

Seeds: All seedsmen list a number of varieties in their catalogue, giving important details about each, such as heavy croppers, very long pods, best for flavour, best for freezing, etc. So, some careful study before ordering is required. (Some varieties do not have the scarlet blooms referred to earlier.)

Soil: A sheltered spot, deep digging and liberal manuring are essential.

Above: A dishful of broad beans ready for use as seeds.

Right: Runner bean 'Crusader'. THOMPSON AND MORGAN

The ideal method is to dig a trench 12 in. (30 cm) deep, line it with at least 3 in. or 4 in. (7.5–10 cm) of manure, old newspapers, rags or grass cuttings and cover with 3 in. (7.5 cm) of good, fine top soil and dress with a good general fertiliser. Leave the top of this trench 4 in. (10 cm) lower than the surrounding soil for ease of watering during the growing period. It will also assist rain to drain down to the roots.

Runner beans will not succeed if careful preparation is not made, particularly in very heavy soils or light sand, or, for that matter in a very cold climate. Cold winds, during the day or night, prevent flowers from setting.

Seed sowing: Sow seeds 2 in. (5 cm) deep and 9 in. (22.5 cm) apart in the trench which should be 12 in. (30 cm) wide, in late April or early May. Place a few extra seeds down the middle for transplanting where any seeds have failed. It is advisable to make three separate sowings at intervals of two to three weeks to extend the harvesting time. The earliest sowing can be made in small individual pots in a warm greenhouse and, after hardening off in the open, the young plants transplanted into their permanent position in the first section of the row.

Whether one, two or three sowings are made they should be in one continuous row, for this helps to give extra strength to the supports – which should be at least 8 ft (2.45 m) high.

Care of the crop

When young plants have grown to 6 in. (15 cm) tall, tie them to supports, after which they will climb naturally and always the same way round – never try to train them in the opposite direction.

Hoe regularly – the plants need all the moisture they can get and weeds will rob them of it. Water regularly every few days and more often in very dry hot weather. It also

helps to give them a liquid feed as soon as the flowers begin to set. When plants reach the top of the supports it is wise to prick out the growing tips. Spray regularly with a very fine spray of water to help pod setting.

Harvesting should continue for about eight to ten weeks, particularly if more than one sowing is made and provided beans are picked regularly and none are allowed to ripen fully. If this is allowed to happen crop will cease promptly.

Pests: *Slugs, millepedes* and the *maggots of bean seed fly*, eat seeds or seedlings. *Aphids*, particularly black fly, infest terminal shoots of both young and established plants. *Capsid bugs* feed on young tissues, causing damaged leaves and bud blindness. *Glasshouse red spider* may infest leaves during excessively dry weather.

Diseases: Brown spots on leaves or stems are usually caused by *anthracnose of dwarf bean* which will go on to cause brown sunken areas on the beans. *Foot and root rot* is caused by fungi and roots will die, stem bases becoming discoloured. *Fusarium wilt* causes leaves to turn yellow and the plants to wilt. Stem bases show red marks. *Grey mould* attacks pods in very wet weather. *Frosts* will kill and cold winds will prevent blooms setting.

Deep Freezing: Runner beans are excellent for deep freezing. Wash well, top, tail and slice. Blanch for three minutes, cool and pack in polythene bags.

Beetroot

There are two main varieties of beetroot – globe and long.

Globe varieties are by far the more popular, partly because they can be sown earlier, are less prone to bolting, and can be grown on a greater variety of soils.

Long varieties should only be grown on deep sandy loam soil and require more attention.

Both varieties are offered by seedsmen as early and late crops and are also available in pelleted seeds.

It is advisable to soak the ordinary seeds overnight in tepid water before sowing to hasten germination, which takes between two and three weeks. Harvesting of the early crops commences ten weeks after sowing; the main crop approximately after sixteen weeks.

Soil: Beds should be thoroughly dug in the late autumn, or early spring, and only truly rotted manure should be added. Peat or other well-rotted compost is equally suitable, but there must certainly be no fresh organic matter.

Beetroot thrives particularly well on light soils, but will produce a good crop on most fertile soils, provided the clay percentage is not high and sufficient compost has been added to avoid cracking or too much drying out during the summer months.

A top dressing of an all-purpose fertiliser should be added about two weeks before sowing.

Sowing Take out 1 in. (2.5 cm) drills in rows 12 in. (30 cm) apart and plant two seeds at 4 in. (10 cm) intervals, rake over and firm down very gently. It is advisable to plant some seeds of both early and main crops, sowing the earlies in March and the main crop in April.

Care of the crop: As seedlings grow, thin out the weaker ones, leaving only one every 4 in. (10 cm). Hoe regularly between rows, but weed by hand near plants, which, if even slightly damaged, will bleed to death.

Regular watering during the summer is necessary to avoid a coarse texture of the root. Mulching between rows with grass cuttings can assist in keeping the bed moist.

During August pull alternate young roots for immediate use, leaving the balance of the crop to mature for later use, bottling or storage.

Above: Beetroot 'Bolthardy'. THOMPSON AND MORGAN
Right: Beetroot 'Little Ball'. SUTTONS SEEDS LIMITED

Harvesting: Beetroot is a very tender crop and damage to the root at the time of harvesting will render them useless. All roots should therefore be lifted carefully in October, by hand, easing the surrounding soil gently with a fork which should be kept well away from the root.

Roots showing the slightest damage should be thrown away, because when cooking they will 'bleed' and not be worth eating. Tops should be carefully twisted or cut off 2 in. (5 cm) from the root.

Storage: Roots can be stored through the winter until March, by being placed in dry peat or sand in a strong box and kept in a frost-proof shed or garage. They are excellent for bottling.

Pests: *Birds* will attack seedlings. *Mangold fly maggots* tunnel through leaves and check growth. *Swift moth caterpillars* attack established plants.

Diseases: *Crown gall* produces large rough growth on roots. *Damping off* causes seedling to die. *Leaf spot* shows brown spots on leaves and is due to various fungi. *Mineral deficiency* in the soil may cause various problems, such as heart rot, roots turning black, and speckled leaves with yellow blotches.

Sprouting Broccoli

Broccoli is one of the brassica family and increasing rapidly in popularity. There are four varieties, all of which should be considered for the vegetable garden because of the widely different harvesting seasons they offer.

Purple sprouting broccoli is the most popular and widely grown variety, but breaks down into three sub-varieties:

★ Christmas purple sprouting – harvested in January and February,
★ Early purple sprouting – harvested in February and March,
★ Late purple sprouting – harvested in April and May.

White sprouting broccoli. This variety produces miniature cauliflower-like spears of excellent flavour during the months of April and May.

Above: Broccoli 'Green Comet'. THOMPSON AND MORGAN

Green sprouting broccoli, frequently referred to as calabrese, produces large green spears from early August until the end of October.

Perennial broccoli is a particularly tall-growing variety and yields only a few small pale green heads in spring and early summer every year. It is useful if there is an unused spot in the garden near a wall or fence, but its crop does not compare with the other three varieties mentioned above.

Each variety grows to a fairly large plant requiring a spacing of at least 2 ft (60 cm) apart and should, where possible, be given a sheltered position out of strong winds.

From the planting of seeds to harvesting Purple and White varieties take approximately forty weeks and the Green variety twelve to fourteen weeks.

Soil: Sprouting broccoli, like all brassicas, prefers a well-drained, heavy to medium soil, which has been well manured the previous October or for a previous crop (but not for any of the brassica family). Prepare the top soil well with a dusting of lime and a general fertiliser one to two weeks before planting.

Sowing: Seeds can be sown in either a seed bed in the open, or in pots or boxes in a warm greenhouse. The advantage of this latter method is that the plants do not have the 'set back' that is normal with transplanting from a seed bed and therefore may make more rapid progress.

If grown in a seed bed in the open, transplanting should be done when seedlings have grown to a height of 5 in. (12.5 cm), but in order to have strong young plants, thinning out will probably be necessary when they are 2 in. (5 cm) tall and transplanting when 5 in. (12.5 cm) tall.

If seeds are grown in a greenhouse, they should be pricked out individually into $2\frac{1}{2}$–in. pots as soon as they are easy to handle (about $1\frac{1}{2}$ in.; 37.5 mm high) using a good potting compost. When these plants have grown to 5 in. (12.5 cm), harden off for a few days and nights in the open and then plant in their permanent position at approximately 2 ft (60 cm) intervals.

It is advisable to water the bed thoroughly the day before transplanting and at the actual time of planting – the hole should be filled with water. Plant firmly at a depth at least 1 in. (2.5 cm) deeper than they were in the seed bed or, if planting from a $2\frac{1}{2}$–in. pot, cover entirely the soil in which they have been growing.

The holes should be given a dusting of the appropriate insecticide against the prevalent pest, cabbage root fly – either calomel dust or 'Bromophos' is suitable.

Crop Protection: Plants should be protected from birds, particularly pigeons, who will, if allowed, strip the leaves down to their veins overnight. Hoe regularly to keep down weeds and keep the soil open to allow rain to seep down easily to the roots. In dry weather water regularly and apply a mulch (grass cuttings will do) to keep roots moist. Spray crop against cater-

Right: Broccoli 'White Sprouting'. THOMPSON AND MORGAN

pillars, white fly and other pests at regular intervals, particularly in August and September. As winter approaches, heel in the plants at the roots and, if necessary, build up soil around stems as an additional support. In very open sites, staking may become necessary.

Harvesting: Take out the centre spear as soon as possible to strengthen the many side shoots that follow. Cut the spears when well formed, about 4 in. (10 cm) long, but before any flower buds have opened. If buds do flower, the crop will cease.

Pests: *Birds*, particularly pigeons – protect with netting. *Maggots of cabbage root fly* which cause plant to collapse. *Flea beetle* eat small holes in leaves. *Caterpillars* of moths and butterflies eat large holes in leaves.

Diseases: All brassicas are subject to the same diseases which include: mildew; club root; damping off; frost damage; leaf spot, and spray damage.

Deep Freezing: Broccoli is excellent for deep freezing. Trim off woody portions of stalk and all outer leaves. Wash and soak in salty water for one hour; grade into sizes. Blanch for three to four minutes, cool and dry. Pack in rigid containers, stalk to tips.

Brussels sprouts

Seeds:
Choose your seeds carefully, bearing in mind the size of your garden, how exposed to wind and gales the site is, whether you require early or late varieties or both, whether you require particularly heavy crops or are more interested in flavour, whether you intend to deep freeze part of the crop. For example, if your garden is small, very open and subject to heavy winds, you will require the dwarf rather than the tall varieties which take up a lot of room and are subject to heavy buffeting so that in no time they have to be staked to stop them lying flat on the ground and spoiling.

Most seedsmen offer hybrid varieties which are increasing rapidly in popularity, because of their compact growth and stems tightly packed with uniform buttons.

The ordinary varieties tend to be the largest croppers, the taller plants yield the larger size buttons.

Soil: The Brussel Sprout is a member of the cabbage family, generally known as brassicas and, like all brassicas, it needs a fertile soil with a liberal amount of organic matter such as cow manure (1½ buckets per square yard/metre) and the addition of a light quantity of lime.

On no account should the soil be allowed to become acid (which the application of lime will prevent). Nor should Brussels sprouts be grown in the same patch which grew brassicas, of any kind, the year before.

Sowing and transplanting: Seeds should be sown between early March and late April. More than one sowing will increase the period of harvesting, particularly if you choose one early and one late variety. Sowing can be done either in a seed bed in the open, or in pots or boxes in the greenhouse.

When grown in a seed bed in the open, transplanting should be done when seedlings have grown to a height of approximately 5 in. (12.5 cm). If seeds are grown in boxes in a greenhouse, they should be pricked out individually into 2½–in. pots as soon as they are easy to handle (about 1½ in.; 37.5 mm) using a good potting compost. Sowings in March will produce an autumn crop. When these plants have grown to 5–6 in. (12.5–15 cm) harden off for a few days and nights in the open and then plant in their permanent position at approximately 2½ ft (75 cm) intervals.

It is advisable to water the bed thoroughly the day before transplanting and at the time of planting – the hole should be filled with water. The holes should then be given a dusting of the appropriate insecticide against the prevalent pest, cabbage root fly – either calomel dust or 'Bromophos' is suitable. Roots should be covered up to the lowest pair of leaves and firmed down into the soil well with fingers or heel.

Crop protection: Plants should be protected from birds, particularly pigeons, who will, if allowed, strip the leaves down to their veins overnight. Spray the crop against caterpillars, white fly and other pests at regular intervals, particularly in August and September. As the weather deteriorates, particularly if the crop is situated in an open position and subject to heavy winds, the taller varieties should be staked to avoid spoilage. Hoe regularly against weeds and keep the soil open to allow rain to seep through to the roots.

Harvesting: Commence harvesting as soon as the buttons have formed into nice tightly formed hard balls, which usually start forming from the base of the stem. Place the thumb over the top of the button and snap downwards, at any one time removing only those buttons that are well formed. All leaves that turn yellow should be removed, also any buttons that have 'blown' or opened. This will happen to a few buttons on almost every stem.

Pests: *Aphids* infest the buttons. *Cabbage white fly* attacks the undersides of leaves at any time. *Maggots of cabbage root fly* feed on the roots and young plants suddenly die. *Flea beetles* cut small circular holes in the foliage of seedlings. *Caterpillars* of butterflies and moths eat large holes in the leaves.

Diseases: *Downy mildew* occurs on seedlings and shows as a white furry fungal growth on the underside of leaves. *Club root* causes roots to thicken and become distorted and plants remain stunted and foliage is discoloured. *Damping off* of seedlings and *wire stem* of mature plants are caused by the

Right: Brussels Sprouts 'Early Button'.
SUTTONS SEEDS LIMITED

fungus *rhizoctonia*. The seedlings collapse at ground level – in older plants the base of the stems become hard, brown and shrunken. In the case of seedlings, overwatering will rapidly cause this problem. *Frost damage* may injure or damage the tissues of plants which have grown soft, and possibly lead to *soft rot* and *grey mould*. *Leaf spot* is due to various fungi, it shows on plants grown too soft, and round brown spots appear on the leaves. *Spray damage* is caused by carelessness when using weed killers and shows as rough wart-like growths at the base of the stem, not unlike club root. Plants remain stunted.

Deep freezing: Brussels sprouts are excellent for deep freezing. Choose small compact buttons only, remove the outer leaves, trim down the stalk and wash thoroughly. Grade into sizes. Blanch for three to four minutes. Cool and drain and pack into polythene bags – the size of bag being determined by the size of the family.

Cabbage

Cabbages are generally regarded as the sheet anchor of the vegetable garden. Greengrocers will tell you that they sell nearly as many cabbages of one variety or another as almost all other vegetables put together.

Seeds: By careful selection of varieties, the gardener need seldom be without, because seeds are available for spring, summer and winter harvests. In each variety there are hosts of seeds to select from in the seedsman's catalogue, each with its own particular peculiarities.

Soil: Like all brassicas, cabbages require well-dug, manured and limed soil, but some period of time should be left between digging and planting. For example, for the spring crop, dig in January – February, and for summer and winter crops in November. All the ground can be dug in October–November if preferred. Before planting commences, rake or rotovate the top 3 in. (7.5 cm) of soil into a fairly fine state and add an all-purpose fertiliser and calomel powder or 'Bromophos' to avoid attacks by cabbage root fly maggots and club root. This work should be completed two weeks before transplanting the young seedlings from the seed nursery.

Above: Cabbage 'Hispi'.
Right: Cabbage. SUTTONS SEEDS LIMITED

Sowing: Take out a trench 6 in. (15 cm) wide and ½ in. (12.5 mm) deep and 'scatter' sow the fine seeds thinly over the area and cover with sifted soil. Firm down with the back of the rake. Germination takes ten to fourteen days.

When seedlings have grown to approximately 4 in. (10 cm) tall, transplant into their permanent site – water the site well the day before planting and water in thoroughly at the time of planting. Give a further dusting of calomel or 'Bromophos'.

Winter varieties – sow outdoors at the end of April or early May and transplant in June.

Summer varieties – sow outdoors in early April and transplant in May or early June.

Spring varieties – sow in July and transplant in September. Certain varieties may differ from this timetable slightly, but instructions are usually printed on the back of the seed packet.

Transplanting: Transplant seedlings into their permanent position when they are approximately 4 in. (10 cm) high and the soil is moist – water in well when planting, giving a further dusting of calomel powder or 'Bromophos' and firm in well with fingers or heel. Give a further watering to help settle the roots in the soil.

Spacing between plants should be between 1½ in. and 2 in. (37.5–50 mm) according to variety, but in the case of spring cabbage, only 6 in. (15 cm) is required between plants, but rows should be 2 ft (60 cm) apart.

Care of the crop: Hoe regularly against weeds and packing of soil.

Spray regularly with insecticide against aphids, caterpillars, flea beetle and other pests. Winds and frosts will loosen soil around stems – firm down as required.

Harvesting: Cut as plants mature. If all three main crops are grown,

cabbages will be available round the calendar.

Pests: *Birds*, particularly pigeons, will, if allowed, strip the leaves overnight right down to their veins. It is advisable to protect the plants with nylon netting. *Cabbage root fly maggots* feed on roots and young plants die. *Flea beetles* make small circular holes in the leaves and eat their way into the hearts of plants.

Diseases: *Club root* causes thickening and distortion of roots. *Damping off* of seedlings is caused by the fungus *rhizoctonia*. *Wire stem* has the same effect on established plants. *Frost damage* may injure the tissues of plants. *Leaf spot* is due to various fungi. *Spray damage* is caused by carelessness when spraying a weed killer.

Carrots

Carrots are roots that do not like fresh manure, or compost, or clay soils. In fact, carrots can be difficult unless conditions are right. There are three main varieties to select from.

The short rooted crop: This variety grows to about $1\frac{1}{2}$ in. (37.5 mm) round and 3–4 in. (7.5–10 cm) long and matures quickly. It is grown as an early crop and should be used immediately, or put in the deep freezer. The carrots have an excellent flavour.

The medium rooted crop: These carrots are generally considered the best all-round crop. When thinning out, the young carrots are ready for immediate use, leaving the balance of the crop to mature for storage, usually in sand, for the winter.

The long rooted crop: This variety has long, tapered, very large roots, and is frequently grown for exhibition. It is more of a problem to grow than the other varieties, because the soil needs very special preparation and deep digging.

Seed: All seedsmen's catalogues list a wide selection from which to choose.

Left: Carrots. SUTTONS SEEDS LIMITED

Above: Carrot 'Freebund'.
THOMPSON AND MORGAN

Soil should be deeply dug and rich, but devoid of clay, if a good crop is to be grown. Many soils will produce satisfactory crops.

Carrots are one of the few vegetables that must not be given manure, which, if applied, tends to cause roots to fork. It is, however, safe to use a plot that was manured the year before.

Sowing: Carrot seeds are very small, so, to ensure a thin sowing and thus avoiding too much thinning out when seeds have germinated, they should be mixed with dry sand or peat before sowing. Sow the early crop in March, and the main crop from April to June.

Make fine drills $\frac{1}{2}$ in. (12.5 mm) deep and 9 in. (22.5 cm) apart, sow thinly and cover with sifted fine soil. Germination will take approximately two weeks and the time from sowing to harvesting is roughly fifteen weeks.

Care of the crop: Thin out the crop as soon as seedlings are large enough to handle, then, as seedlings grow, thin out once or twice more – these later trimmings can be eaten.

According to variety, the remaining carrots require 3–6 in. (7.5–15 cm) spaces to grow to full size. If the soil is dry always water before thinning out but firm down the soil around the remaining crop. This will help to protect the roots from carrot root fly and keep them moist. Hoe regularly to keep weeds down to the minimum and handweed near roots. Keep the soil permanently moist by gentle watering in dry weather. If the soil is allowed to dry out at all, either watering or rain will then cause roots to split.

Protect the crop against carrot root fly by application of insecticide.

Harvesting: Pull carrots as and when required from July onwards and lift the main crop in mid-October for storage in a dry place under sand or peat. Remove all soil from the carrots and cutting off all the leaves. The crop will remain in good condition until the following March.

Pests: *Aphids* infest leaves. *Maggots of carrot fly* tunnelling into the roots is indicated by a red discoloration of the foliage.

Diseases: *Damping off* will cause seedlings to turn yellow and die. *Motley dwarf disease* is a virus disease that turns the foliage red. *Sclerotinia disease* can develop both during growth and in storage. The roots are destroyed by a white fluffy fungus. *Soft root* is due to a bacteria. It occurs in storage if roots have been injured or if storage

arrangements are too damp. *Violet root rot* shows as a violet fungus over roots.

Deep freezing: Carrots are good for deep freezing, but, as indicated above, can be stored equally well provided the necessary conditions are observed.

Cauliflower

Cauliflower, like cabbage, is a brassica and, by the selection of various varieties, is available for harvesting virtually all the year round.

Seeds: The frequency of harvesting that is required, and the particular months preferred, will control your selection of seeds.

The summer varieties mature during June, July and August and seeds may be selected to mature for each of these months. Seed sowing should commence in January in a heated greenhouse, or outdoors in early April. It is advisable to select two different varieties because some mature more quickly than others.

The autumn varieties mature during September, October and November. Again there is a wide selection from which to choose – either the large variety or the dwarf Australian strain. Sowing begins in April and continues until the end of June.

The winter varieties mature between January and May. Seed sowing should be made between April and July, according to the variety selected.

Soil: As for all brassicas, soil preparation is essential to ensure a good crop. Dig the bed in the autumn and dose heavily with cow or horse manure, and a spreading of lime to ensure that no acidity is left in the soil. Two weeks before seedlings are transplanted, prepare the top 3 in. (7.5 cm) of soil with a rake or rotovator, adding a dressing of a general fertiliser and a dusting of calomel or 'Bromophos' against cabbage root fly maggots and club root.

Left: Winter cauliflower. SUTTONS SEEDS LIMITED

Sowing: Take out a 6 in. (15 cm) wide trench ½ in. (12.5 mm) deep and 'scatter' seed lightly and cover with fine sifted soil. Firm down with the back of a rake.

Transplanting: When the seedlings are 4 in. (10 cm) high lift carefully, retaining as much soil around the root as possible. The bed to which they are to be transplanted should be well watered the day before. The seedlings should be watered into their holes thoroughly, and a dusting of calomel or 'Bromophos' added to the hole. Plant the seedlings to the same level as they were in the seed bed. Give a further watering the same or following day to help settle the roots in the soil.

Care of the crop: Give protection against birds, particularly pigeons, and occasionally spray with insecticide against caterpillars and aphids. Feed with a general fertiliser occasionally. Water regularly and never allow plants to become too dry or they will mature too early and have very small heads. Spring and summer varieties require protection from the sun by covering the heads with a large leaf. Similarly, the winter varieties heads should be protected from frost and snow by the same method.

Harvesting: Do not allow the heads to become too large before cutting. Cut early in the day, except during frosty weather. If it becomes necessary to cut earlier than required for consumption, put the head in a sealed polythene bag in the bottom drawer of the refrigerator, or pull up the root and hang it upside down in a dry shed, where it will keep perfectly well for up to two weeks.

Pests: *Birds,* particularly pigeons, will, if allowed, strip the leaves overnight right down to their veins. It is advisable to protect with nylon netting. *Cabbage root fly maggots* feed on roots and young plants die. *Flea beetles* make small circular holes in the leaves and eat their way into the hearts of plants.

Diseases: *Club root* causes thickening and distortion of roots. *Damping off* of seedlings is caused by the fungus *rhizoctonia. Wire stem* has the same effect on established plants. *Frost* may injure the tissues of plants.

Deep freezing: Cauliflowers are excellent for deep freezing. Choose firm young cauliflowers. Wash well and break into florets. Blanch for 3 minutes in acidulated water (i.e. add 1 teaspoonful (5 ml) of lemon juice or vinegar to each ½ pint (300 ml) of water used), cool and drain. Pack in polythene bags.

Above: Celery 'Golden Self Blanching'. THOMPSON AND MORGAN

Celery

Self-blanching varieties. These varieties are unquestionably the best for an amateur to grow. They obviate the necessity of trenching or blanching which is both time consuming and very hard work. The self-blanching varieties are crisper, but milder than the trench varieties.

Trench varieties are difficult to grow and make hard work, but they do have a better flavour, although they can be stringy. Only gardeners with a rich and deep top soil should consider growing this variety.

Seed: Choose with care and be sure to select only seeds treated to prevent leaf disease. Some seedsmen, who sell treated seed, give a *medical warning that seeds thus treated must not be used for flavouring soups or medicinal purposes.*

Soil: Choose a sunny position in the garden. For self-blanching varieties dig thoroughly at the end of March or early April, and apply 1½ buckets per square yard/metre of manure. Trench varieties require to have a trench 12 in. (30 cm) deep and lined with a generous quantity of manure covered by 6 in. (15 cm) of soil. Give a dressing of a general fertiliser for both varieties.

Sowing: Sow seeds in April in boxes or pots in seed compost under glass. Prick out when ¾ in. (20 mm) high into boxes of potting compost. Keep the plants growing steadily until ready for transplanting into the permanent beds.

Transplanting: Transplant the seedlings into permanent beds in late June or early July. Self-blanching varieties should be 6 in. (15 cm) apart in squares, and watered in well. Trench varieties should be planted 9 in. (22.5 cm) apart in the trench which should be 15 in. (38 cm) wide and as long as the number of plants require. The trench should be thoroughly dowsed with water after planting. Spray the soil in both cases with 'Bordeaux Powder' as a protection against leaf disease.

Care of the crop: Watch the crop for brown blisters (celery fly). Pick off the affected area and spray with 'Malathion'. Apply slug bait or pellets around the plants or in the trench. Water heavily, particularly in dry weather, and give the crop regular doses of liquid feed.

Trench varieties, when about 10 in. (25.5 cm) high, should be tied loosely and gradually earthed up as they grow. Spray the crops again with 'Bordeaux Powder' from August onwards. In September, complete the earthing up, leaving only the foliage tops showing.

Harvesting: Self-blanching varieties should be lifted as required, but *very* heavy frosts may affect these varieties adversely.

Trench varieties should be ready for harvesting by mid-November, approximately two months after they were finally earthed up. Hard frost is unlikely to affect these varieties – in fact, it improves the flavour.

Pests: *Maggots of carrot fly* may damage roots and stem. *Celery fly* tunnel into the leaves, causing blotches and stunting the growth of the plant. *Slugs* feed on the stalks and kill plants.

Diseases: *Arabis mosaic virus* and *cucumber virus* cause stunted and distorted growth and may be seen as yellow and green mottling. *Celery heart rot* causes the centre of plants to turn into a slimy brown mass. *Celery leaf spot* shows on leaves and stalks as small brown spots. *Damping off* – fungi cause seedlings to droop and die.

Chicory

Chicory provides an excellent salad from December to March – a time when lettuce is not readily available. It has a slightly bitter taste when eaten, but is crisp, like celery, and extremely pleasant when eaten with cold meats.

Soil: Rich, but not recently manured, soil gives the best results. The ground used for some other previous vegetable which required manure is ideal.

Sowing: Take out a drill ½ in. (12.5 mm) deep in the permanent position and sow thinly in April or early May. Cover the seeds with sifted soil and firm down with the back of the rake. As the young seedlings grow thin out, so that eventually the plants remaining have a distance of 9 in. (22.5 cm) between them.

Care of the crop: Chicory is generally a trouble free crop provided the bed is hoed regularly to keep down weeds. Give the bed a light dressing of an all-purpose fertiliser and keep the plants generally damp during dry weather. When the green tops begin to die down at the end of October, lift and clean the soil from the roots. Cut the foliage away close to the crown and trim the thin bottom end off the root.

The roots should then be stored in moist peat or sand in deep boxes, in a horizontal position. Cover the crowns with a 12 in. (30 cm) layer of sand and store where the temperature will remain above freezing point – in the garage or a dry frost-proof shed where temperatures will remain between 40° and 45°F (4.5–7°C). To encourage a few roots at a time to produce their 'chicous' move batches into a higher temperature, in perhaps the boiler house, where the temperature does not exceed 55° to 60°F (13–15.5°C) and the roots will soon produce their chicous, but do avoid too much heat. These roots must also be kept in soil or sand and darkness. It is a good idea to use two 12–in. pots – one to contain the roots, the other to stand on top into which the chicous will grow.

Very little watering is required provided the peat is moist when the roots are put in. However, if the plants show signs of flagging, give them a modest watering.

Absolute darkness is required for the growth of the chicous, and burying the roots in peat and covering with sand should ensure this. When the white chicous have grown to about 6 in. (15 cm) long, slice off from the root. They are then ready for eating.

Pests and diseases: Chicory is generally trouble free, except that the roots may be attacked by the *caterpillars of swift moths* or *slugs.*

Cucumbers

Growing in a greenhouse

The plants strictly for those with a heated greenhouse, require a minimum temperature of 60°F (15.5°C), or for the all-female varieties 70°F (21°C).

Few gardeners would elect to grow these varieties of cucumber unless they had spare room in the greenhouse in which they were growing tomatoes and it is on that assumption that the following advice is given.

Right: Chicory.

Above: Cucumber 'Topsy'.

Select either the ordinary variety or the all-female variety – the latter have the advantage that they grow only female flowers, thus avoiding the task of removing male flowers, and give a prolific crop; but they produce smaller cucumbers, and require more heat. Consequently they are more expensive to grow.

Soil: Special soil is required made up of 2 parts loam, 1 part manure or compost, plus a good dosage of bone meal. Alternatively, a special compost can be purchased which has been prepared for the purpose. If this is used, it is better to grow each plant individually in 12–in. pots.

Sowing: Sow seed in good seed compost ½ in. (12.5 mm) deep in 2½–in or 3–in. pots during March. Keep in a warm temperature of over 60°F (15.5°C) or for all-female varieties at 70°F (21°C). Plant on into special soil, or in 12–in. pots when four leaves have fully formed.

Care of the crop: All male flowers must be removed as they appear. Female flowers have a tiny cucumber behind the bloom; the male, a thin stork.

Stems should be trained up a wire, cane or trellis. Prick out the terminal shoot when the roof of the greenhouse is reached. Side shoots should be trained horizontally, keeping them well away from the glass. The soil should be kept slightly moist: the humidity required is similar to that required by tomatoes. Good ventilation is necessary during the day. Feed with liquid fertiliser as soon as the fruit begins to swell. Spray or dust occasionally against aphids with 'Malathion'.

Harvesting: Cut as soon as the fruits have reached a reasonable size. Cropping ceases if the fruit is left on the vine too long.

Growing in a cold greenhouse or cold frame

The same procedures as above are necessary but the whole programme should be delayed. The seeds will have to be germinated in similar warmth (that can easily be done in a boiler house indoors), but should not be transplanted into soil or 12–in. pots until early June, which means seeds should be sown around 20 April to 1 May. Frames or greenhouses should be shut at night against cold air.

Ridge cucumbers

There are three main varieties, all of which can be grown in the open.

Ordinary varieties, which are thick and medium sized with a knobbly surface.

Hybrid varieties. This variety gives a better quality fruit, is less acid and does not cause indigestion. Suttons seeds have a variety known by the delightful name of 'Burpless Tasty Green' which makes the point perfectly.

Japanese variety, which are the longest and smoothest skinned of all outdoor varieties.

Seeds: Seeds should be soaked overnight in tepid water before sowing.

Sowing: Seeds germinate within ten days provided they are kept in a temperature of 60°F (15.5°C) – another possibility for the boiler house. Sow well separated 1 in. (2.5 cm) deep in good seed compost, in May, and select the strongest seedlings for planting on into 5–in. pots, when they have produced four leaves. It is a good idea to plant only one or two seeds in 2½–in. pots rather than in boxes, keeping the stronger of the two for development.

Transplanting: When plants have grown to 4 in. (10 cm) tall, harden off outdoors if they have been protected in a cool greenhouse, cold frame, cloche, or even by an upturned jam jar, and plant outdoors in early June 2½–3 in. (6.25–7.5 cm) apart. Even then, it is wise to cover at night for the first week.

Ridge cucumbers need a good sunny position, well protected from strong winds, because even this variety of cucumber is not particularly hardy. The soil must be well drained and very rich in manure.

Only a few plants will be required, so rather than prepare a whole bed, just make small individual beds for each plant.

Care of the crop: When seven leaves have developed, pinch out the growing tip to encourage the growth of lateral shoots. These can be allowed to remain on the ground or trained up nylon netting, which must be well supported. Though it is necessary to keep the soil moist, do not water the actual plant, but give the leaves a fine spray in dry weather. Once fruit starts to swell protect the soil from the sun by placing black polythene around the root and surrounding area, or mulch with grass cuttings.

Contrary to greenhouse cucumbers, do *not* remove male flowers, otherwise the bees will not do their job.

Harvesting: Cut the fruit before it is too large to encourage further fruiting. The size of crop will depend on how early is the first autumn frost, but in normal seasons a good crop can be expected during September and October.

Pests: Greenhouse varieties may be infested by *glasshouse red spider mites. Woodlice* cut holes in the leaves of both indoor and outdoor varieties.

Diseases: Grey mould, mosaic virus, root rot and powdery mildew are not uncommon.

Above: Curled kale 'Frosted'.

Kale

This is another of the brassica family of vegetables, but by no means so demanding in its cultural requirements.

It is perhaps less popular than other brassicas in the south of England, but in areas where the weather is more extreme it is one of the main winter crops. There are four main species of kale:

Curly kale is by far the most popular and is very widely grown in the north of England and Scotland. Each leaf has a particularly curly edge resembling parsley in appearance.

Plain leafed kale is a very tall species and much coarser than curly kale. It gives a prolific crop and is very hardy.

The rape kale is the summer kale and gives young tender shoots in the spring and early summer.

Leaf and spear kale produce young leaf shoots in the spring, which are followed by spears resembling broccoli spears.

There is also *Borecole:* A packet of

seed will provide a most interesting and varied crop – *but not for eating.* It is a hardy plant and provides a very pleasant colour for winter flower arrangements in the home. It has a frilled foliage in shades of purple and purplish green and the colours greatly improve after the first frost. Arranged in a floating bowl, borecole is shown off to its greatest advantage.

Seeds: The various species are listed in most seedsman's catalogues. Kale is regarded as a highly desirable vegetable, particularly by Sutton & Sons who claim it is 'the best green winter vegetable'.

Soil: Kale is far less difficult in its soil requirements than other brassicas such as cabbage and, particularly, cauliflower, and it will produce successful crops in nearly any type of soil. The soil does not have to be dug and prepared in any particular way – merely rake the top 2 in. (5 cm), remove all weeds, add a little general fertiliser and plant out and firm the soil down hard.

Sowing: Take out a 6 in. (15 cm) trench ½ in. (12.5 mm) deep, scatter seeds lightly, cover with sifted soil and firm down with the back of a rake. Sow the early varieties in April and later varieties in May. Rape kale should be sown in its permanent bed and not transplanted.

When seedlings are 6 in. (15 cm) tall, transplant them into their permanent bed leaving 2 ft (60 cm) between plants – a particularly suitable place is a site left vacant after lifting early potatoes. Water the bed well the day before planting and water in thoroughly at time of planting. The time between sowing and harvesting for early varieties is thirty weeks, and for late varieties thirty-six weeks.

Care of the crop: Hoe regularly to control weeds but always firm down afterwards to prevent plants swaying in the wind. Spray against aphids such as white fly and caterpillars of butterflies and moths. Pay particular attention to the soil around stems and firm up regularly.

Harvesting: Leaves and shoots should be picked when very young to obtain the best flavour. The harvesting of kale is the exact opposite from the harvesting of Brussels sprouts. Start at the crown of the plant, taking off a few young leaves at a time and always removing all yellowing leaves at the same time. The removal of leaves will encourage the growth of side shoots which, in turn, should be gathered young. If left too long there will be a considerable loss of flavour and the kale could become bitter. Like Brussels sprouts the flavour is greatly improved after the first frost.

Pests: *Caterpillars of butterflies and moths* eat holes in leaves. *Flea beetle* eats small circular holes in the leaves of young seedlings.

Diseases: *Damping off* causes seedlings to collapse and die. *Wire stem* is caused by the fungus *rhizoctonia* affecting the base of the stem. *Violet root rot* attacks roots under the ground.

Leeks

There are three main varieties of leeks, but a fairly wide selection of each variety is available.

The earlies. This variety must be sown in a warm greenhouse and transplanted in the open in April.

Mid-season is sown in the open March–April and transplanted in June.

Late varieties are sown in the open April–May and transplanted in July.

Soil: Leeks are easily grown in any well-drained garden soil which should have been well dug and lightly manured the previous autumn, or, alternatively, manured for a previous crop. The ground should not be too beaten down from winter rains or waterlogged.

Sowing: Sow very thinly in a seed bed and when 6 in. (15 cm) high transplant the seedlings into their growing position. There is an advantage in using pellet form leek seeds when planting in a warm greenhouse. The bed to which the seedlings are being transplanted should be well watered the day before to ensure that the soil is thoroughly moist.

Plant 6 in. (15 cm) apart in rows 9–12 in. (22.5–30 cm) apart by making a hole 6 in. (15 cm) deep with a dibber at least 1½ in. (37.5 mm) in diameter and simply drop the seedling in and fill the hole with water to settle the roots. Do not on any account fill the hole with soil, this will be done gradually and naturally by rainfall or gentle watering as time goes by.

Care of the crop: Hoe regularly to keep down weeds. Give at least two light dressings of an all-purpose fertiliser to improve the quality of the crop. To increase the length of blanched stem, draw up soil around the stems in September or October. Keep well watered at all times during dry weather.

Harvesting: Commence harvesting in early November by lifting with a fork. The leeks may be allowed to remain in the soil during the winter and dug as required or, if the land is needed for other purposes, they may be raised in November and covered with earth in some convenient corner of the garden. By using the early, mid-season and late varieties, leeks are available from September to March inclusive.

Pests: *Onion fly maggots* may attack roots and infest the bases of plants when they die.

Diseases: *White rot* causes leaves to turn yellow and die back; the roots to rot, and the bases of plants to become covered with white fluffy fungal growth.

Lettuce

With care and very simple planning, there is no justifiable reason why lettuce should not be available from April to November, though the long harvesting will mean using various varieties which are suitable for the

Right: Leeks 'Lyon-Prizetaker'.

Above: Two fine varieties of lettuce. SUTTONS SEEDS LIMITED

spring, summer and autumn seasons. The necessary information is to be found in all seedsmen's catalogues.

It is perhaps unwise to give particular advice as to variety, because people's tastes vary enormously, but one variety in particular 'Little Gem' does seem to give particularly good results at all seasons and is also renowned for excellent flavour and crispness. It is a small cos lettuce growing to about 8 in. (20 cm) tall and of an upright nature, thus it takes up only a small amount of room per plant – plants need only be 6 in. (15 cm) apart. This variety develops a good heart and, unlike most other cos lettuce, does not require tying to keep its shape.

There are four main varieties from which to choose:

Butterheads are very popular, largely because they are quick to mature, but the leaves are soft and smooth and are generally summer varieties. Some varieties are recommended for spring crops and for forcing under glass.

Crispheads produce large hearts of curled crisp leaves. They do well sown in a cold frame for an early crop, otherwise they are generally a summer variety. Being large plants they require space and should be planted 12 in. (30 cm) apart.

The cos lettuces are tall and upright. The larger varieties require tying as they grow to keep them erect. Leaves are very crisp and of excellent flavour.

Loose-leaf varieties do not produce hearts. The leaves are curled and should be picked a few at a time, leaving the root in the ground to continue growing.

Soil: Well-manured soil is essential for growing good lettuce. The bed should be dug to about 8 in. (20 cm) in depth and dosed with 1½ buckets of well-rotted manure per square yard (square metre) and given a dressing of lime to avoid any acidity. The soil should be kept damp at all times.

Before each sowing rake thoroughly 2–3 in. (5–7.5 cm) deep to form very fine soil, add a dressing of all-purpose fertiliser and a small quantity of pest killer, such as 'Bromophos' or Murphy's combined Pest and Disease Dust. Having raked all well in together, take out a small drill with a hoe ½ in. (12.5 mm) deep ready for sowing.

Sowing: Sow seeds very thinly and for only about 6 ft (1.80 m) at a time, covering with very fine soil, which should give about forty seedlings seven to ten days later. Do this once every ten to fourteen days throughout the entire sowing time from March to September to be assured of a continuous supply of lettuce ready to harvest.

Thin out seedlings as soon as they are easy to handle, the thinner the sowing of seed the less work of this nature there is to do. Continue thinning as the seedlings grow, keeping the strongest plants. Eventually most varieties will require 12 in. (30 cm) distances, but smaller varieties like 'Little Gem' require to be only 6 in. (15 cm) apart. The seedlings that are thinned out when 3 in. (7.5 cm) high should be transplanted, but enormous care is required not to sever the root at its end and to keep as much soil clinging to the fibres as possible. Water in very thoroughly and keep well watered until the transplant has established itself in its new home.

Care of the crop: As soon as seedlings show through the soil, put down an ample dressing of slug pellets or bait. Lettuces appear to be the favourite food for slugs and nothing is more annoying than to be told that a magnificent specimen is 'full of slugs'. Give further supplies of slug bait as the lettuce matures. Watch for green fly and spray immediately any signs are seen, otherwise the crop will be lost.

Harvesting: The time between sowing and harvesting varies according to variety – anything from eight to fourteen weeks. Lettuce should be harvested as soon as a good firm heart (that, is for crisp head and cos) has formed and before the centre leaves start pushing their way upwards. Place your hands around the plant and press: if it is good and firm it is ready for harvesting. If left, even for a few days, lettuces will bolt and be useless, but you can pick several at a time and keep them for a few

Right: A good example of loose-leaf lettuce. SUTTONS SEEDS LIMITED

days in the refrigerator crisper (bottom drawer) provided they are stored in a sealed polythene bag. Butterheads should be harvested where the plants have reached a good size. Pick in the mornings while the heads are damp.

Pests: *Slugs* and *snails* are particularly troublesome. Watch also for *green* and *white fly*, and *cutworms* and other caterpillars.

Diseases: Be alert to *damping off, grey mould,* and *virus diseases. Drought* and underwatering will cause early bolting.

Marrows and Courgettes

There are two main varieties of marrow – the bush and the trailing varieties. The trailing varieties require plenty of room to spread, but the bush variety can be grown in a relatively confined area.

Soil: The soil must be very rich in humus and well drained. Large quantities of manure should be added to all soils, and the lighter the soil the more manure is required.

Sowing: For the trailing variety it helps to sow on a tall mound and for each plant to be trained to spread in a different direction away from the centre. The bush variety may be sown singly in pockets in convenient places in the vegetable garden, but for both varieties a sunny protected position away from cold winds is advisable.

Seeds should be soaked in slightly tepid water overnight and planted 1 in. (2.5 cm) deep in the prepared positions – make sure the manure extends to at least 12 in. (30 cm) deep. Sow outdoors in May or in a greenhouse in individual pots at the end of April. Sprinkle a general fertiliser over the planted area and scatter a generous amount of slug bait. If a greenhouse is available, plant seeds individually in 3–in. pots using a good seed compost and when 4 in. (10 cm) tall, harden seedlings off in the open and plant out in the prepared beds in June.

Left: Marrows. SUTTONS SEEDS LIMITED

Above: Courgette 'Golden'. THOMPSON AND MORGAN

Care of the crop: Continue applying slug bait. When the marrows start to swell, feed with a general-purpose fertiliser every two weeks. Keep the soil moist at all times and water thoroughly around the roots regularly, giving two gallons of water to each plant on most days in dry weather. Trailing varieties should have the tops of their main shoots pinched out when 4 ft (1.20 m) long.

If very few bees are about, help pollination by folding back the petals of a male flower and push it gently into two or three female flowers (tiny marrows behind flowers).

Harvesting: When the marrows reach 10 in. (25.5 cm) in length they should be cut from the stem with a sharp knife. Regular cropping is essential if fruiting is to continue.

Storing: Provided cutting is done regularly, up to twenty-four marrows per plant can be harvested. Leave the last three or four fruits on the plants until October when they will be fully grown and ripened. Cut these and store them in an airy frost-proof shed, if possible suspended in a net.

Pests: *Slugs* are the chief enemy.

Diseases: *Virus* causes stunting of plants and distortion of leaves. *Grey mould* is caused by too much watering and turns the fruit yellow. *Powdery mildew* is a fungal growth on leaves.

Courgettes

Courgettes are a dwarf bush variety of the marrow. They grow very quickly and give a most prolific crop.

Fruits should be cut when only 4 in. to 6 in. (10–15 cm) long. Plants should be examined daily to avoid allowing fruits to grow too large, when, if permitted, they would considerably reduce the yield.

In every sense, the method of growing is identically the same as for marrows.

Mushrooms

Mushrooms can be grown virtually all the year round indoors and during the summer and early autumn outdoors. The first and most essential thing is to obtain the correct compost. Ideally, there should be an equal mixture of horse droppings from a healthy horse that is mainly fed on corn and oats, and straw from the floor of the stable. Horse droppings gathered separately and mixed with fresh straw will not do. It is possible to use cow manure and straw, but the mushrooms grown on this compost will be poor. The mixture of horse manure and straw will generate enormous heat and should be turned regularly until it has cooled to 70°F (21°C).

The ideal compost is, of course, out of the reach of many people who live nowhere near a suitable stables. There are excellent alternatives. It is possible to buy a ready-made compost prepared specially for mushroom growing, or to buy some bales of wheat straw and to treat them with a proprietary chemical manufactured for the purpose. These chemicals contain ammonia nitrate, ammonia sulphate, lime, ammonia phosphate, ammonia carbonate and other compounds mixed to the right proportions. The results from the ready-made compost or chemically treated wheat straw are entirely satisfactory.

Preparing mushroom beds: When the mushroom compost is ready for use it should be made into beds – flat indoors, in ridges outdoors. If indoor space is limited, large boxes containing compost can be used.

Flat beds for indoor growing should be packed down firmly to a depth of 10-12 in. (25.5-30 cm) by 4 ft (120 cm) wide, and as long as the room permits.

Ridge beds for outdoor crops should be 2 ft 6 in. (76 cm) wide at the base, and tapering to 6 in. (15 cm) at the top, by 2 ft 6 in. (76 cm) high. When these beds are made it is essential for them to be on flat, well-drained ground. The same site must not be used in successive years – in fact, the same site must not be used again for three years.

The best site is close to a wall or fence to give protection against winds or draughts, and away from constantly dripping water. If necessary, wind-breaks should be erected. Cold frames are excellent for growing mushrooms because they provide their own protection against wind and rain. Other suitable places for growing mushrooms include greenhouses, polythene tunnels, garden sheds, garages and cellars, with a flat bed in each case.

How to plant: Mushroom spawn should be inserted 2 in. (5 cm) deep at 18 in. (45 cm) intervals. About ten days later the beds should be examined closely and if the spawn has become fluffy and bluish-grey in colour, and shows tiny threadlike roots entering the compost, the bed is ready to be cased with a pure subsoil, which may be mixed with peat if desired. To case the beds the subsoil should be laid about 1 in. (2.5 cm) deep on flat beds, and 2 in. (5 cm) thick at the bottom tapering to 1 in. (2.5 cm) at the top of ridge beds.

After casing the bed completely, cover those in the open or in cold sheds all over with 8-10 in. (20-25.5 cm) of straw to keep in the heat.

Care of the crop: Beds should be kept on the dry side because if they are too damp the spawn will be killed, but a light watering from a watering can with a rose should be given every seven to ten days; mix a tablespoonful (20 ml) of salt with each gallon (4.5 l) of water. Indoors, the growing place should be sprayed with a solution of water and Jeyes Fluid, but the mixture must not be allowed to fall onto the growing beds.

Mushrooms will begin to appear after six to eight weeks indoors and eight to ten weeks outdoors and may be expected to continue for three months.

In greenhouses, cold frames, sheds etc mushrooms require good ventilation, a temperature of 50-60°F (10-15.5°C) and a humid atmosphere. The glazing of greenhouses and cold frames should be sprayed with whitewash mixed with clay or soot to keep out the light and so help to maintain an even temperature. In both cases some ventilation is essential, but it must not be allowed to create a draught.

Pests: *Maggots* of various mushroom flies tunnel into the stalks and caps.

Diseases: Unhygienic conditions will cause diseases to develop.

Above: Mustard and cress.

Mustard and Cress

These salad vegetables are usually grown together, normally under glass or in a warm spot – even in the kitchen. Cress takes about three days longer to mature than mustard, and should be sown so much earlier to keep pace with its companion. These vegetables are so easy to grow that they are often grown by children on damp blotting paper.

Sowing: Seeds may be sown in very fine soil or on any material that will retain moisture, such as blotting paper, pads of cotton wool, sacking or rags. These latter are better than soil as no soil sticks to the seedlings.

Spread the seeds thickly and evenly on the top of the growing medium and press them down gently, but do not cover. Place the medium with its seeds in a container, such as a tray, and cover that with black polythene or brown paper, and put it where there is a temperature of approximately 50°F (10°C). When the seeds have germinated remove the cover and keep the medium moist with tepid water.

Harvesting: Mustard will be ready for harvesting in ten to twelve days and cress in thirteen to fifteen days. There is absolutely no reason why a regular supply of these two green salads should not be available all the year round if they are grown in a reasonable temperature.

Pests and diseases: If mustard and cress are grown on anything but soil no pests or diseases should be encountered.

Onions and Shallots

Growing from seed

In selecting seeds from catalogues, decide whether strong or mild flavours are required and whether for normal home consumption or exhibition.

The bulb varieties are available either flat or globe shaped and are grown for their large size. They also store well, hung in a dry shed, for winter use. When being thinned out and still small, they can be used for shades. Some varieties are especially suitable for bottling.

The salad varieties are grown specifically for salads, are white skinned and milder than the bulb varieties.

Seeds: Bulb varieties should be sown in March or April, according to the climate, to produce crops from July onwards. *Exhibition* bulbs should be sown from mid to late August for a June harvest the following year.

Salad varieties, which are very quick growing, should be sown in succession from March to September to ensure a regular and plentiful supply throughout the summer and autumn.

Soil: Dig thoroughly in the late autumn, adding a generous amount of manure (1½ buckets per square yard/metre). When seed sowing

Some excellent bulb onions.
SUTTONS SEEDS LIMITED

time arrives, rake or rotovate the top 3 in. (7.5 cm) of soil, adding a small amount of general fertiliser at the same time.

Sowing: Take out ½ in. (12.5 cm) deep drills 9 in. (22.5 cm) apart and cover with sifted soil and firm down with the back of a rake. Thin out bulb varieties as they grow, first to 2 in. (5 cm) apart and later to 6 in. (15 cm) apart, (the thinned out baby onions maybe used for salads). At this stage treat soil with soil pest killer against onion fly. Keep the bed moist at all times.

The salad varieties should be treated in the same manner, but pulled as required.

Care of the crop: Weed by hand around seedlings and hoe carefully between rows to keep weeds down. Provided the soil is dressed against onion fly, kept reasonably moist and given an occasional liquid feed, little else needs to be done.

If flower stems appear they should be snapped off, but they are a clear indication that the bed is becoming too dry.

Harvesting: When the crop has matured, the foliage will buckle and fall over. Three weeks later the onions should be carefully lifted and left on the soil to dry for three or four days. Then tie them together in small bunches and hang them up in a dry airy place for storage and use during winter.

Pests: *Maggots of onion fly* tunnel into the bases of young plants and the bulbs of mature plants. *Stem and bulb eelworms* attack leaf and bulb tissues, which become distorted.

Diseases: *Downy mildew* turns leaves grey. *Onion smut* shows on leaves of young plants as grey blister-like stripes. *Soft rot* causes stored onions to become soft and slimy and have a particularly unpleasant odour. Careless storing is usually the reason. *Storage rot* is due to blue mould or similar fungi – and careless storing. *Virus disease* shows as yellow streaks at the base of leaves, which crinkle. *White rot* causes leaves to turn yellow, roots to rot, and bases of bulbs to become covered with a fluffy growth.

Growing onions from onion sets

All seedsmen offer onion sets, which are immature bulbs which have been specially grown and are ready for planting in April. Within days of planting they start to grow.

There are many advantages to growing from sets as distinct from seed (except for salad onions or *exhibition onions*). They save a lot of work and avoid most of the possibilities of crop failure, for instance they are not attacked by onion fly maggots or mildew.

The ground should be prepared in a similar manner as for seeds but the soil does not have to be of such a fine texture. The sets should be put into the ground so that the soil covers all but their tips, and firmed well down. Watch for birds pecking them from the ground, which they frequently do, and just replace.

Shallots

Shallots are a useful addition to the vegetable garden, take up very little room and are very useful in the kitchen for flavouring in stews and garnishing. Their flavour is similar to, but milder than the onion.

Buy the shallots and plant them in the early spring, in a similar way to onion sets, where they will multiply and reproduce themselves approximately ten times, giving ten shallots where one was planted.

Lift in July, separate and allow to dry. Store in bags.

Parsnips

There are three main varieties of parsnips, the short-rooted, the intermediate and the long-rooted varieties.

Soil: Only gardeners who have fine light soil should attempt to grow the long-rooted variety, in soil that has been well dug and manured for some previous crop. Almost any soil, however, will yield a good crop of the other two varieties, but if the soil is very heavy then the short-rooted variety should be chosen. In flavour there is little or nothing to choose between any variety.

The gardener with only a small area available for vegetables should bear in mind that parsnips occupy the ground for a very long period – almost twelve months – and thus restrict the growing of other and perhaps more popular vegetables.

Parsnips require that beds are dug deeply, but *fresh* manure or compost should not be added. The area should be given a light dressing of lime and a general fertiliser.

Sowing: Ideally, seeds should be sown in February but, in practice, the soil is seldom in a fit condition for sowing to take place until March or even April. The soil should be broken down and well raked so that the top 3 in. (7.5 cm) is a fine tilth, then take out drills ½ in. (12.5 mm) deep and sow in batches of four or five seeds at 8 in. (20 cm) intervals and cover with sifted soil. Make sure the soil is moist, if necessary by giving a gentle watering. When the seeds have germinated, thin out the seedlings, leaving only the strongest plant of each batch to grow.

Care of the crop: Hoe regularly to keep down weeds. Otherwise, very little attention is required, except to watch for any pests or diseases that may attack the crop.

Harvesting: Parsnips are usually left in the ground until they are wanted, particularly as the flavour improves the longer they are left – but not later than February. If the ground is wanted for other purposes they can be lifted at the end of November, trimmed by cutting off the tops close to the crown, stored in a shed or a sheltered position and covered with sand.

In any event, it is wise to lift a portion of the crop in the late autumn so that some are available when the ground is frozen hard or covered with snow.

Pests: *Maggots of celery fly* bore holes in the leaves, causing them to shrivel. *Earwigs* feed on the leaves.

Diseases: *Parsnip canker* shows as reddy-brown or black lesions on the top of roots. *Splitting root* is caused by lack of moisture. *Forked root* is caused by stones, lumpy soil or fresh manure being present in the soil.

Garden peas

The three main varieties of garden peas are:

The wrinkled variety, so called because the seeds are wrinkled. This is the heaviest cropping, sweetest and largest variety, and by far the most widely grown.

The round variety because of their smooth round seeds, which are the hardiest variety and may be sown as early as November for a crop in the following May.

The Mangetout variety which are especially suitable for eating the whole pods while peas are still very small, in much the same way as runner beans.

Sowing dates are in November, and again from March to the end of June. Seedsmen offer early, mid-season and late varieties which vary in height from 1½ ft to 4 ft (45 cm–120 cm). Always use round varieties for early crops.

Soil: The principle of rotating crops is very important with garden peas – the ground must have a two-year rest in between crops if a good yield is to be obtained.

The ground should be well dug in the autumn with a generous application of manure. Two to four weeks before sowing, the top 6 in. (15 cm) should be well prepared with additional manure and a general fertiliser and well raked together – this is a crop where a rotovator is of particular value to ensure that the soil, manure and

Right: Choice garden peas. COLOUR LIBRARY INTERNATIONAL

fertiliser are broken down together into a fine mix to give the seeds a maximum opportunity for germination and the growing of a strong crop. Ensure that the soil is not acid by adding a dusting of lime.

Sowing: Take out drills in the prepared bed 6–8 in. (15–20 cm) wide and 2 in. (5 cm) deep and scatter seeds evenly throughout the drill. It is wise to be generous with the seed, because in many gardens some are likely to be eaten by mice. After sowing, firm down, covering the soil lightly with the back of the rake. Germination takes from seven to twelve days, according to climatic conditions.

Care of the crop: Spray ten days after flowering to protect against maggots. Hoe regularly to keep down weeds. When seedling are approximately 3 in. (7.5 cm) tall, stick with small twiggy sticks for support and two weeks later stake again with 3 ft (90 cm) stakes. Tall varieties will need longer stakes. Keep the beds moist at all times and water well in dry and hot weather. Protect the crop against birds by covering with netting as soon as pods begin to form.

Harvesting: Pods are ready for picking when well filled – apply a little pressure with thumb and forefinger as a check. Pick the peas regularly to ensure a continuing crop. The approximate time from sowing to harvesting is fourteen weeks (except for November sowing).

Pests: *Aphids* infest young shoots and leaves. *Millepedes* attack germinating seeds. *Pea moth caterpillars* eat into pods and feed off peas. *Pea thrips* can appear in large numbers and cause silvering of pods and leaves.

Diseases: *Damping off* may cause November sowings to rot. *Downy mildew* causes grey furry patches on the underside of leaves. *Grey mould* attacks both pods and stems when the weather is very wet. *Powdery mildew* is a white coating on leaves and stems. *Manganese deficiency* will affect seeds. *Virus diseases* cause dead patches in the foliage and distortion of leaves.

Deep freezing: Peas are one of the best vegetables for the deep freezer. Shell only young peas, blanch for one minute, cool in cold water, dry, and pack in polythene bags.

Above: Pepper 'Gold Topaz'. THOMPSON AND MORGAN
Right: Pepper (Capsicum annum).

Peppers (Capsicum)

The culinary varieties may be grown outdoors in the warmer parts of Britain, but they will fruit earlier and do better if grown in the warmth of a greenhouse.

There are a number of species and among the culinary species the hardy *capsicum annium* and *capsicum frutescens* are well worth trying. *Capsicum annium* is a shrubby perennial species with mid-green oblong leaves. The several varieties of this species differ only in the shape of the fruits and are red, yellow or green in colour. *Capsicum frutescens* is a bushy perennial with oblong, slender, pointed leaves. The fruits are bluntly conical.

If a warm greenhouse is not available, it will be necessary to buy young plants from a nursery or garden centre because, basically, peppers are a tropical plant like the tomato and will not survive without the necessary warmth.

Other species of peppers are available for growing the fruits of chilli, cayenne and paprika.

Soil: Peppers require a fertile, well-drained soil which has been well dug and to which well-rotted compost or manure has been added.

Sowing: Sow seeds in a good seed compost during March in a

temperature of 60°F (15.5°C). When the seedlings are large enough to handle, prick out into individual 3–in. pots, using a good potting compost. If the plants are to be grown outdoors, harden off during the day-time and continue to protect at night until all fear of frost is past. Plant out in late May or early June.

If the peppers are to be grown in a greenhouse they should be transplanted from 3–in. pots on to 9–in. or 12–in. pots. On hot days stand them out in the open, returning them to the greenhouse at night.

As only a few plants are likely to be required, the potting method is probably best. When standing them outside, put the pots in a protected position where they will receive the maximum sunshine.

Pests: *Glasshouse red spider* mites attack leaves, and the *Capsid bug* distorts growing plants and leaves. Otherwise, peppers are generally free from disease.

Potatoes

There are three main classes of potatoes – the Early, Second Early and Main Crop.

The Early is an uneconomic crop to grow, but it is especially pleasant to eat. It is uneconomic because it is usually dug very early before the tubers have been allowed to swell to their full size, at the end of June or early July. This class does not store well.

Second Earlies, or Early Main Crop, may be dug at the end of July or early August for eating up to the end of November.

The Main Crop, which should be dug by the end of September, not only bear the heaviest crops, but store well, and in many cases improve with storage.

Thus, the Second Earlies provide potatoes from July to November and the Main Crop provide potatoes through the winter to the following June when the early potato becomes available again.

Grow only as many Earlies as are needed for immediate eating.

Always buy 'immune' or 'certified' varieties which are guaranteed against wart disease.

Soil: The first thing to remember is that potatoes do not like a lime soil, but they will grow well in almost any type of soil. It is said that anyone developing land for the future should first grow a crop of potatoes which will help clean the land. Start by digging the land in the autumn, adding a good quantity of manure, compost and peat, and leaving the soil in large lumps for the frosts to break down during the winter. Before planting, rake the top soil thoroughly down to at least 3 in. (7.5 cm) or, if available, use a rotovator to loosen the top soil down to 6 in. (15 cm) mixing, in a good dressing of 'Bromophos' or similar insecticide to prevent attacks from wire-worm.

Planting: As soon as your seed potatoes are received from the seedsman, set them out carefully on trays or in seed boxes which should be lined with 1 in. (2.5 cm) of peat. Keep them in a light position, but away from the rays of the sun and away from any possibility of frost, and allow the shoots to grow to 1 in. (2.5 cm) in length. Allow only two shoots to remain at planting time for Earlies and four for the Second Earlies and Main Crop. The ideal size for a seed potato is about that of a chicken's egg, but if they are larger they may be cut into two or more pieces.

Earlies should be planted as soon as soil conditions will allow – late March or early April. Second Earlies should be planted in mid-April and the Main Crop in late April.

Take out drills 5 in. (12.5 cm) deep and set the seed potatoes – Earlies and Mid-Earlies 12 in. (30 cm) apart and Main Crop 15 in. (38 cm) apart. Replace the soil gently so as not to damage shoots, using particularly the fine soil that has been raked or rotovated, and build a small ridge along the line of planted potatoes. The rows for Earlies and Mid-Earlies should be 24 in. (60 cm) apart and the Main Crop 30 in. (76 cm) apart.

Care of the crop: As soon as shoots appear above the soil, commence 'earthing up' to protect against frosts; at first only a little soil should be drawn up over them. As the crop grows continue this process so that ultimately the ridge is at least 6 in. (15 cm) high. Make sure that all weeds are removed at regular intervals and, provided the soil between rows is kept clean and fine, it is a simple matter to draw the soil up with the use of a rake. Keep the entire bed moist at all times and in a drought use a hose and give a good dosing of water between the rows. In wet summers watch for blight which can affect the Main Crop.

Harvesting: In mid-June examine one root of the Earlies and, provided the potatoes are large enough – the size of a golf ball or slightly larger – start harvesting. Insert a fork well away from the root to avoid damaging the potatoes and lift forward into the trench. Mid-Earlies should be treated in the same way, making the examination in mid-July.

The Main Crop should be left until the tops die down, when they should be removed and the potatoes left in the ground for one to two weeks before digging. The approximate timing for this is mid-September.

Before storing, the potatoes must be allowed to dry out to avoid trouble during storage time. If the weather is dry, they may be left where they are dug; if not, they should be taken into a dry and airy shed and set out on the floor to dry.

If in doubt as to whether the Main Crop is ready for harvesting, rub the skin of a potato hard with the thumb and, provided the skin is well set and does not easily rub off, it is ready.

Store in a wooden box or sack (not plastic, which will make them sweat) and keep in a frost-proof shed.

Left: Potato 'Pentland Hawk'.

Pests: *Aphids* may affect seed potatoes and growing plants. *Potato cyst eelworm* may infest the roots and cause minute brown cysts. *Slugs* and *wire-worm* eat holes in the growing tubers.

Diseases: *Common scab* causes raised scabs with ragged edges on the tubers. *Dry rot* may affect seed potatoes. *Gangrene* affects stored potatoes, which will rot. *Rust spot* shows as scattered brown marks on the flesh of the potato. *Potato blight* causes brown blotches on the leaves, which will turn black and rot.

Radishes

The three main varieties of radishes are Round, Intermediate and Long. There are also winter varieties, but they are not generally popular. The secret of growing good radishes is to ensure that they grow quickly so that they do not become tough and fibrous.

Soil: Radishes enjoy a fertile, well-drained soil, but they will usually grow well anywhere. Where the soil is not up to standard add peat or well-rotted compost, and an all-purpose fertiliser which should be raked into the fine top soil.

Sowing: Sow thinly $\frac{1}{2}$ in. (12.5 mm) deep in rows 6 in. (15 cm) apart but do not sow too many at a time; better a few and often. Cover the seeds with sifted soil and firm down. Start sowing in January under cloches and continue at fortnightly intervals to ensure a continuous supply.

Care of the crop: Summer varieties do not require to be thinned, but plants should be 1 in. (2.5 cm) apart. Hoe to keep down weeds, but weed by hand near seedlings. Protect the crop against birds, and watch for flea beetle. Winter varieties should be thinned out to 6 in. (15 cm) apart.

Harvesting: Do not allow the crop to get old and overgrown, when the radishes will become inedible – too hot, hard and fibrous. Harvesting from seeds sown in January and onwards should continue from May until September. Winter varieties may be left in the ground until required from October onwards.

Above: Radish 'Scarlet Globe'.
Right: Perpetual spinach.

Pests: *Birds* peck holes in the roots. *Codling moth* and *apple sawfly larvae* burrow holes into the cores. *Aphids, woolly aphids* and *caterpillars* infest the foliage and *flea beetles* perforate leaves. *Capsid bugs* pierce the roots.

Diseases: *Apple mildew* and *brown rot.*

Spinach

Spinach is regarded as one of the more nutritious of green vegetables and has the great merit of being suitable for harvesting all the year round. There are four main varieties: the Summer; the Winter; the New Zealand, and Perpetual Spinach.

The Summer varieties have round seeds and grow very quickly provided the soil is properly prepared, and can be harvested within three months of sowing. The disadvantage of this variety is the speed at which it runs to seed in warm dry weather.

The Winter varieties generally have prickly seeds and make a valuable winter crop to be harvested from October until April.

The New Zealand variety is not a true spinach, but thrives well on very poor soil where true spinach will not grow. Like Perpetual Spinach it is milder in flavour and frequently more popular with children than the true spinach.

Perpetual Spinach is, in fact, a

variety of beetroot grown specially for its leaves and is generally much more favoured than the New Zealand spinach. Seeds sown in April will yield crops for harvesting in July and onwards until the next May. It has a mild flavour and grows well on poorer types of soil, and is more popular with children. This variety does equally well in sun or frost.

Soil: Good rich soil with plenty of manure or compost is particularly desirable for both the Summer and Winter varieties. Summer varieties, in particular, will bolt very rapidly unless the soil is properly prepared and kept damp. Digging, which should be done during the winter, should be deep, and large quantities of humus and a good dusting of lime should be added. Two weeks before sowing, give an application of a general fertiliser. These general soil requirements apply only to the Summer and Winter varieties.

New Zealand and Perpetual Spinach are far less demanding and will grow well even in poor or sandy soil. Application of lime and general purpose fertiliser are also advised for all varieties before sowing.

Sowing: Sow seeds thinly 1 in. (2.5 cm) deep in rows 12 in. (30 cm) apart, watering the soil well before sowing. Cover with fine soil and firm down with the back of the rake. Be generous in the number of plants grown, otherwise there is a tendency to over-pick the leaves when harvesting, which will kill the plants.

Summer Spinach: sow every two weeks from March to the end of June.

Winter Spinach: sow in August and again in September.

Perpetual Spinach: sow in April.

Care of the crop: Thin out seedlings to 3 in. (7.5 cm) apart and ultimately to 6 in. (15 cm) apart, keeping the strongest young plants to mature. Water regularly and flood during very hot weather. Winter spinach is affected by frost, so it should be covered by cloches from the end of October. Hoe regularly to keep weeds under control.

Harvesting: Spinach and spinach substitutes are all gathered in the same way. Begin by eating the seedlings of the second thinning, eating the whole plant except for the roots. When plants have matured, start by taking the outer leaves first, but do not strip one plant – allow as many leaves to remain as are pulled. Be sure to commence harvesting while the largest leaves are still young and tender.

Take leaves only, not the stems, by picking them off with thumb and fingers without tearing the main stem. If this is found to be difficult, use scissors.

Summer Spinach: harvest from June to September.

Winter Spinach: harvest from October to April.

Perpetual Spinach: harvest from July to May.

Pests: *Mangold fly* maggots may tunnel leaves. *Millepedes* may attack seeds. Perpetual Spinach is generally trouble free.

Diseases: Perpetual Spinach is normally trouble free, but Summer and Winter Spinach can be affected by *damping off, downy mildew, leaf spot* and *cucumber virus.*

Swedes

The swede is, in fact, a variety of turnip, but is hardier and will frequently succeed where difficulty has been experienced with ordinary turnips. It is milder and sweeter and rarely gets 'woody'. There are two varieties, Purple Top and Bronze Top.

Soil: There is no need to dig the land, provided it was well dug and manured for some previous crop. Rake the surface well and add a light dressing of lime and a general fertiliser.

Seeds: Having prepared the bed, leave it to settle for about four weeks, then take out drills ½ in. (12.5 mm) deep and sow seeds very thinly, and cover with sifted soil which should be firmed down with the back of the rake. Drills should be 15 in. (38 cm) apart. The sowing should be done between May and June.

Care of the crop: As the seedlings grow, thin out in easy stages until they are 9 in. (22.5 cm) apart. Be alert for turnip flea beetles, which in some areas exist in large numbers. Keep the soil well hoed against weeds and be sure that the soil is moist at all times to avoid root splitting. Apply a liquid feed occasionally.

Harvesting: If the ground is not required for other crops, swedes may be left in the soil over winter. If storing is necessary, the roots may be lifted at the end of November or early December, the leaves cut off just above the crown, and stored in dry peat or sand in a box which should be kept in a dry shed and away from frost.

Pests: *Turnip flea beetle* – a minute black insect which attacks leaves of seedlings, particularly in dry weather.

Diseases: *Club root* causes roots to become swollen. *Damping off* causes seedlings to collapse at soil level. *Soil deficiency* can cause grey or brown patches on the lower half of the root. *Mildew* sometimes appears on the leaves of seedlings. *Soft rot* occurs if the crop is carelessly stored. *Splitting root* is caused by irregular watering.

Sweet Corn

It is now quite easy to grow sweet corn in all parts of Britain provided the 'F. hybrid' varieties which are now offered by all leading seedsmen are selected.

If sweet corn is a great favourite with the family, there is a selection available of both early and late maturing plants.

The 'open-pollination' varieties are more difficult to grow but produce heavier crops than the hybrids.

Soil: The bed in which the sweet corn is to grow should be dug at

Right: Swede 'Ne plus ultra'.

least one spit deep, adding well-rotted manure or compost (fresh manure must not be used) with the addition of a dressing of peat. Choose a well-sheltered position away from winds. Rake the top 3 in. (7.5 cm) into a fine tilth and add a light dressing of an all-purpose fertiliser.

Sowing: The seeds are large and therefore very simple to sow. Take out a drill 1 in. (2.5 cm) deep and 24 in. (60 cm) apart in squares or rectangles, rather than in single rows, so as to assist wind pollination of the flowers. It is advisable to sow under cloches, except in the warmest parts of Britain, and even then provided it is not a cold spring.

The earliest time to sow is early May, and sowing may continue up to the first week in June. If early sowing is required, and a greenhouse is available, it is wise to plant seeds individually in 3–in. pots, using a good seed compost such as John Innes or Levington. The seedlings can then be planted into their permanent position once the dangers of frost have passed, after hardening off during the daytime for a few days. When planting out, leave 18 in. (45 cm) between each plant and cover the roots to just below the bottom leaves.

Care of the crop: Sweet corn is a very shallow-rooted crop and roots will appear at the base of the stem as the plants grow. These must be covered regularly with soil or well-rotted compost, ultimately earthing up to 8 in. (20 cm). Water well except in wet weather. When cobs begin to swell, liquid feeding is necessary.

Remove side shoots as they grow. Hoe regularly, but weed around the plants only by hand for fear of damaging shallow roots. Plants should be staked when 3 ft (90 cm) high.

Left: Sweet corn.

Harvesting: It is not always easy to determine when the cobs are ripe. One method is to watch for the appearance of the silks (long silky hairs) and then allow three weeks for the ripening of the cobs. In the early stages the grains in the cobs are watery; they then become thick and creamy and ready for harvesting. If the liquid is thick and doughy the cob is too ripe and unfit to eat. To test for ripeness, pull back part of the sheath and squeeze one or two grains between fingernail and thumbnail.

When ripe, you can either snap the cob from the stem or cut.

Each plant will produce one or two cobs.

Pests and diseases: Birds attack the tips of cobs. Otherwise sweet-corn is trouble free.

Deep freezing: Unquestionably, it is best to cook immediately for the best flavour, but sweet corn does deep freeze well. Freeze only fresh young cobs. Remove husks and silks, blanch for four minutes, cool, dry and pack individually in waxed paper or tin foil.

Tomatoes

Grown in the greenhouse

There are numerous cultural methods that have been developed over the years, almost entirely because tomato plants are so susceptible to disease that the utmost care must be exercised to ensure the use of a disease-free soil. The amateur, growing only enough tomatoes for his own consumption, will be wise if he buys soil or soil-free compost such as John Innes, Levington or Fisons.

Sowing: For the professional grower there are four main sowing seasons:

★ In the last week of June or early July for a December crop,
★ In mid-August for a February crop,
★ In October for an April crop,
★ In December for a crop in June.

For the amateur, sowing may begin as early as mid-November if a very early crop is desired, but that does mean keeping the greenhouse at temperatures as high as 60–65 °F (15.5–18.5 °C) to ensure success.

For the average grower a reasonable timetable to follow is: sow in seed trays or 6–in. pots using John Innes or Levington seed compost. If you have a particularly warm place in the house at, say, 60 °F (15.5 °C), but not higher than 65 °F (18.5 °C), such as a boiler house, it is helpful to place the seed trays or pots there to speed up germination – that is, of course, if the greenhouse is not heated to the high, and now very expensive, 60 °F (15.5 °C).

Fill the trays or pots with slightly warmed compost, pressed down to ½ in. (12.5 mm) from the top. Then sow the seed thinly, cover lightly, and press down gently to keep the seed below compost level. Cover with paper or dark polythene until germination has taken place. The seedlings can then be moved from the warm place into the greenhouse, where a temperature of 55 °F (13 °C) should be maintained – though anything less than 60 °F (15.5 °C) may delay the date of harvesting by a couple of weeks.

Care of the crop: Ten to fourteen days after sowing, the seedlings should have developed their first two leaves. They should then be potted on individually into 4–4½-in. pots of clay, plastic, peat (Jiffy pots) or polythene – the last are by far the least expensive and very suitable for the purpose. It is wise to use a prepared compost, either John Innes No 1 or Levington soil-less compost.

It is important to treat the seedlings with extreme care when potting on, always holding the leaf tips and not the stem. Fill the pots with compost to within ½ in. (12.5 mm) of the top and use a dibber to make holes that are large enough to take the seedling without restricting the roots, then firm in gently by finger and give the pot a sharp tap on the potting bench. Finally, water slightly, preferably with a fine spray. This should be repeated at regular intervals, but the frequency will vary according to the amount of sunshine in January, February and

March. *At no time should the young seedlings be overwatered* because nothing kills young plants more quickly than saturation, especially when a soil-less compost is used.

Make sure that all windows are kept clean so that the plants enjoy the maximum amount of daylight and sunshine.

As the young plants grow in the greenhouse (which should be kept as near as possible at 60°F; 15.5°C) constant attention to detail is essential. Adverse conditions to look for as the plants grow are:

Symptom	Cause	Remedy
Pale green leaves	Lack of light – too high temperature	Lower temperature
Very dark green leaves – shrivelled or curled leaves	Too much feeding being given	Reduce feeding and water well
Blue coloration of leaves and stunted plants	Greenhouse too cold	Increase heating and check for draughts
Specially long growth	Too much heat or insufficient lighting	Spray leaves with water, reduce heat and move plants to a lighter position in greenhouse
Poor growth and yellowing leaves	Too much water	Reduce watering and give a good feed
Leaf mottling	Unbalanced feed	Fertiliser might be old stock
Wilting, distortion	Virus or fungal diseases	Possibly curable, but better to buy new plants and start again

Regular feeding of the plants is essential, *once the first truss has started to grow* but not before. Various proprietary feeds are available, such as 'Tomarite', and clear instructions as to their use accompany them.

When the plants have grown to healthy and strong plants about 6 in. (15 cm) tall, (probably by the end of March) they should be potted on into 9–in. or 12–in. pots. Numerous methods are employed by professional nurserymen, but for the amateur there are three main methods.

The first is to use 9–in. or, preferably, 12–in. pots of clay, whalehide, polythene or plastic, and John Innes No 3 or Levington potting compost. Ensure good drainage by placing a generous layer of 'crocks' (broken pots, brick or stone) at the bottom of the pots. If John Innes is used, leave $\frac{1}{2}$ in. (12.5 mm) between the compost and the rim of the pot for ample watering. If Levington is used, when watering should be lighter but more frequent, fill the pot to the brim to discourage overwatering. Plants must be kept in the greenhouse, standing on boards, bricks or ashes, in a steady temperature of 60°F (15.5°C). The pots should be placed 2 ft 6 in. (76 cm) apart to allow for full growth. At the time of potting

Above: Tomato 'Pixie'. THOMPSON AND MORGAN
Right: Tomato 'Grenadier'. SUTTONS SEEDS LIMITED

canes should be put into position ready for tying as the plants grow – early caning prevents the damage to roots that later caning might cause.

The second method is to use the soil in the greenhouse, but care must be taken to ensure that it is good healthy soil, disease free and well drained.

The plants should be set out 2 ft 6 in. (76 cm) apart with a good measure of manure or peat under each plant, and caned at the same time. Keep the greenhouse at 60°F (15.5°C).

The third method, which is gaining rapidly in popularity, is by 'ring culture'. When young plants are repotted into 9–in. or 10–in. pots receptacles should be bottomless and contain 8 in. (20 cm) of John Innes No 3 or Levington compost. The receptacles should then be placed on a bed of weathered ash or clinkers, or a mixture of 3-parts gravel and 1-part coarse vermiculite. Tomatoes grown by this method develop fibrous feeding roots in the bottomless pots and the roots that provide liquid grow down into the gravel below, which must be kept permanently moist. Once the first truss has set, the container should be fed with liquid manure every three days.

In all methods, the plants will continue to require careful attention and regular feeding and watering (an average of 2 pints (1.125 litres) per plant per week, but more in sunny weather).

As soon as the first blooms are fully in flower on the first truss they should be sprayed with 'Tomato Set' to assist the setting of the fruit, and this should be repeated ten days later. Regular tying of the main stem to the cane as it grows is essential to take the strain of the weight of fruit, which is considerable – 6–10 lb (2.700–4.500 kg) per plant.

Side shoots *must* be removed as they grow – these are the small shoots that grow between the main leaves and the main stem. This should be done with a sterilised sharp knife and on no account by the nails of thumb and finger – and even more especially not by smokers.

Canes, referred to earlier, are very expensive to buy. An alternative, and far less costly method of securing the plants as they grow, is to run a stout wire along the top of the greenhouse, either longways or across, at 2 ft 6 in. (76 cm) intervals and tie string to the wire, with the other end secured to the pot, and tie the main stem to the string as it grows.

Harvesting: Tomatoes should be harvested most carefully. Always snap the stem at the small joint near the fruit, leaving the calyx on the tomato. Do not leave fruit, when picked, in the sunshine.

The amount of fruit harvested varies, but 6–10 lb (2.700–4.500 kg) should be obtained from a plant. It is advisable to stop the plant when six trusses have formed, to ensure good-sized fruit.

Pests: Tomatoes are very prone to attack by aphids – *white, green* and *black fly*. A wise precaution is to spray fortnightly with Murphy's greenhouse aerosol or liquid Malathion, or, if you prefer, mufane Lindane smoke. When the season is over, the greenhouse should be thoroughly cleansed and fumigated – prevention is much easier than cure.

Grown outdoors

The tomato, growers should remember, is a tropical plant that has been developed and is now entirely a hybrid. A fact of life is that only under glass and in warmth will the tomato grow satisfactorily, especially in its young life. The tomato stands no chance at all of growing in the open until the soil is warm and frost long past. It is a very tender plant.

Sowing and planting: No attempt should be made to grow tomatoes outdoors from seed unless the proper conditions exist. The purchase of strong healthy young plants from a nurseryman or garden centre is the course to follow, but not before the end of May or in early June in colder regions.

In any district it is essential to choose a sunny site and arrange the rows of plants so that they receive the maximum amount of sunshine. No tomato will ripen if planted in a bed with a northern aspect, and beds facing east are likely to suffer from cold winds.

In small gardens there is no reason why tomatoes should not be grown in 9–in. or 12–in. pots, provided a good potting compost is used. In pots, too, it is a simple matter to give the plants protection if the weather turns cold during a spell of the north or north-east winds that so often develop during dry spells in the British summer.

Before the young plants are bought, the beds should already have been properly prepared. Dig well and break down the soil, add peat, strawy manure and, where necessary, a good dressing of lime. A few days before planting, add a dressing of a suitable fertiliser and rake it well into the top soil. Hoof and horn may also be added. Where the soil is light the amount of organic material to be dug in should be considerably greater than in heavy soils – about 2 bucketsful per square yard/metre. All work should be completed in April to allow the various ingredients to perform their functions in readiness for the plants.

Care of the crop: At the time of planting it is advisable to give the young plants some protection with, say, large flower pots or boxes at night, unless the night temperatures are above normal.

Planting distances should be 2 ft (60 cm) between plants and 2 ft 6 in. (76 cm) between rows. Water the plants only if the weather is warm and dry, and even then only slightly – they are best left without water until their roots have found their way into the new soil, but it is essential not to disturb the soil in

Right: Tomatoes in the greenhouse.
SUTTONS SEEDS LIMITED

which they have been growing in their pots. As you plant, put in the stakes so that there is no fear of damaging the roots later. Side shoots should be removed with a sterilised knife (do not use your fingernails) as they grow and the plant should be stopped by pinching out the top once five trusses have formed.

If the weather stays dry for the first ten days after planting, watering should be started and then continued regularly. Mulching around the roots is beneficial, but a small area must be left for watering and feeding. In July spray with Burgundy mixture against disease.

When the flowers appear it is helpful to spray them with 'Tomato Set' and feeding with liquid manure should begin as soon as the first truss has formed its small tomatoes. Feeding should continue regularly, twice every week.

If the growth of leaves is excessive some may be removed to let the sun through to the fruit. Discoloured leaves should all be removed.

At the end of the season green fruit can be ripened by being put in a warm, dark place such as a drawer in the kitchen. Alternatively, green tomatoes can be used to make excellent chutney.

Turnips

There are two main varieties of turnip: *The Earlies*, which are quick growing and cannot be stored – so a few only should be sown at a time – and should be eaten as harvested. *The Main Crop*, which can be stored and used throughout the winter. A great advantage of most varieties of this crop is that the green tops are also edible and regarded as the most nutritious of all green vegetables.

Soil: Turnips differ from almost all other root crops in soil and growing requirements. They need a rich, well-manured soil, in which they must be made to grow quickly. They should mature in eight to ten weeks from the sowing date. The Early varieties require the soil to be well dug with plenty of manure and to be allowed to settle for about a month before sowing. The Main Crop does not require the land to be dug, merely raked over from a previous crop that was well manured: but a light dressing of lime and general fertiliser should be added in the top 3 in. (7.5 cm) of soil.

Sowing: There are two main sowings – from the end of March to the end of June for The Earlies, which should be sown a few at a time, and in July and August for the Main Crop. Crops grown especially for their green tops should be sown in August–September. Seeds should be sown very thinly ½ in. (12.5 mm) deep, covered with fine soil and firmed down with the back of the rake. If the soil is not moist, a gentle watering will be necessary.

Care of the crop: Thin out as seedlings grow, a few at a time, until The Earlies are 6 in. (15 cm) apart. Do the same for the Main Crop until turnips are 12 in. (30 cm) apart. Roots grown for their green tops do not need to be thinned out. Cut their tops when 5 in. (12.5 cm) long. The soil must be kept moist at all times to avoid root splitting. Feed with liquid manure occasionally.

Harvesting: Early turnips should be pulled before they become too large – when they are about 2½ in. (62.5 mm) thick. The Main Crop may be left in the ground until required, unless the autumn is exceptionally wet, when it is wise to lift by mid-November. Remove the leaves just above the crown and store the roots in dry peat or sand in a box which should be kept in a dry shed. The crop should keep well in these conditions until March.

Pests: *Flea beetle* – a minute black insect which bores small holes in the leaves, particularly in dry weather.

Diseases: *Club root* causes roots to become swollen. *Damping off* causes seedlings to collapse at soil level. *Soil deficiency* can cause grey or brown patches on the lower half of the root. *Mildew* sometimes appears on the leaves of seedlings. *Soft rot* occurs if a crop is carelessly stored. *Splitting root* is caused by irregular watering.

Left: Turnips 'Model White'.

Pests and diseases

The gardener's enemies

Young plants have a number of enemies – late frosts, birds (pigeons in particular), and numerous pests in the soil and air such as cabbage-root fly, wireworm, chafer grubs, catworms and carrot fly. It is therefore advisable to cover beds with nylon netting against birds and place a sprinkling of an appropriate insecticide, such as 'Bromophos' or calomel dust in holes when planting (particularly brassicas).

These and other problems are not confined to young plants and a close study of the Pests and Diseases chart will greatly help to reduce these problems. It cannot be emphasised too strongly that as soon as trouble is spotted, action must be taken to cure it. Frequently it is wiser to use pest controls before the problems arise, on the basis that prevention is better than cure.

In the section of this book that deals with individual vegetables, the author has listed the various pests and diseases which are known to affect each plant – a list he admits he has taken from an encyclopaedia, for he himself has been fortunate and has not by any means experienced all these misfortunes, although he has had first-hand experience of quite a number of them: hence the advice that 'prevention is better than cure' – a policy he himself has adopted with excellent results in the past few years.

Snail. ALL ILLUSTRATIONS BY MURPHY CHEMICALS LTD

Pest or disease	Symptoms and where found	Treatment
Aphid (greenfly, black fly) Capsid	Clustered on leaves and stems Tattered holes in leaves, brown spot on pods	Spray Systemic Insecticide, Liquid Malathion, Lindex, Fentro or Liquid Derris when pests or damage seen. Or dust with Malathion or Combined Pest & Disease Dust.
Grey mould	Grey fungus	Spray Systemic Fungicide during flowering or use Combined Pest & Disease Dust.
Leaf miner	Whitish blisters in leaves in May/June	Spray Systemic Insecticide or Liquid Malathion when mines first seen.
Caterpillars	Eat holes out of leaves in spring and summer	Spray Fentro or Liquid Derris or dust with Sevin, Derris or Combined Dust.
Root fly	Maggots feed on roots, plants fail to establish	Dip roots in Colomel Dust when planting, or water in with Lindex.
White fly	Tiny white moth-like insects underneath leaves	Spray undersides of leaves with Liquid Malathion and repeat twice at 7 day intervals.
Aphid	Leaves turn red, plants are stunted	Spray Systemic Insecticide, Liquid Malathion, Fentro or Liquid Derris.
Carrot fly	Maggots mine in roots	Treat seed with Combined Seed Dressing. Dust rows lightly with Gamma-BHC Dust.
Leaf miner	Mines in leaves	As for beetroot.
Leaf spot	Brown rusty spots on leaves and stems	Spray Bordeaux Powder every 2 to 3 weeks or use Combined Pest & Disease Dust.
Red Spider	Leaves stippled with pale spots, especially in hot, dry summers	Spray Systemic Insecticide, Liquid Malathion or Liquid Derris or Combined Pest & Disease Spray.
Root aphid	Grey, mealy pests on roots, plants wilt	Water soil with Liquid Malathion.
Grey mould	Plants rot off at soil level	Spray base of seedlings and plants with Systemic Fungicide.
Onion fly	Whitish maggots in onions	Dust Calomel into seed bed and dip root in paste of Calomel Dust.
White rot	White mould on base of onion at harvest	
Pea moth	Maggots found in the pods	Spray Fentro during flowering.
Blight	Brown marks on leaves. Haulm collapses	Spray Bordeaux Powder from early July at 3 week intervals or use Combined Pest & Disease Dust.
Flea beetle	Small round holes in seedling leaves	Treat seed with Combined Seed Dressing or dust seedlings with Sevin, Derris or Combined Dust.
Cutworms and Leatherjackets	Feed on stems and roots, especially just at soil level	Dust soil with Sevin Dust.
Slugs and Snails	Feed on all plant parts	Water plants and soil with Slugit Liquid or scatter Slugit Pellets.
Wireworms and Millepedes	Feed on seeds and roots	Treat seed with Combined Seed Dressing and dust soil with Gamma BHC Dust (not carrots or potatoes).

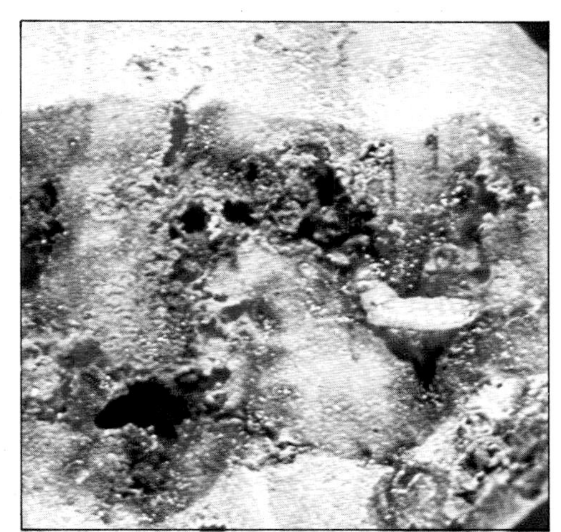

Carrot fly.

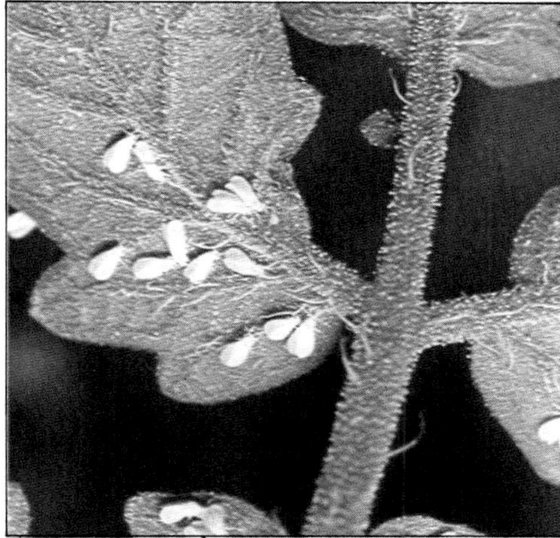

White fly (tomato).

Cabbage root aphid.

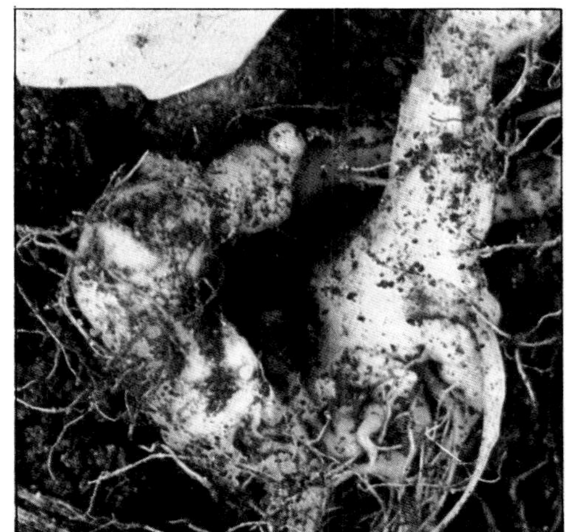

Club root.

Wireworm.

Millepedes.

Bordeaux Powder (Murphy). Controls blight on potatoes and tomatoes, and celery leaf spot.
Breakthrough (Murphy). The seaweed soil conditioner. Improves soil structure, provides essential trace elements.
Calomel Dust (Murphy). Controls club root, onion white rot, onion fly, cabbage root fly.
Combined Pest and Disease Dust (Murphy). Controls aphids, caterpillars, white fly, mildew, blight, etc.
Combined Pest and Disease Spray (Murphy). Aerosol for pest and disease control on outdoor lettuce, tomatoes and cucumbers.
Combined Seed Dressing (Murphy). Gives increased germination and healthier plants by controlling pests and diseases of seedlings.
Derris Dust (Murphy). Controls flea beetles and young caterpillars.
Derris Liquid (Murphy). Controls aphids (green fly and black fly, caterpillars, etc.).
Fentro (Murphy). Controls aphids, caterpillars, capsids, etc.
Foliar Feed (Murphy). Formerly 'FF'. Feeds through leaves and roots to stimulate growth and increase crop weight.
Gamma – BHC Dust (Murphy). Controls wireworms, millepedes, root flies, and other soil pests.
Lindex Garden Spray (Murphy). Controls aphids, cabbage root fly and numerous other pests.
Malathion Dust (Murphy). Controls aphids, leaf miner, white fly, etc.
Malathion Liquid (Murphy). Controls aphids, leaf miner, white fly and numerous other pests.
Ramrod© (Murphy). Prevents annual weeds in cabbages, cauliflower, broccoli, kale, sprouts, and in onion and leeks.
Sevin© Dust (Murphy). Controls caterpillars, flea beetle, cutworms, leatherjackets, etc.
Slugit Liquid (Murphy). Controls slugs and snails walking over treated soil or feeding on treated plants.
Slugit Pellets (Murphy). Controls slugs and snails. Mould resistant and showerproof.
Systemic Fungicide (Murphy). Controls grey mould, powdery mildews and many other diseases.
Systemic Insecticide (Murphy). Controls aphids, capsid, leaf miner, and many other pests.

Acknowledgements
We wish to thank Murphy Chemicals Ltd., Wheathamstead, St. Albans, Herts. for the information contained in this chart.

Crop failure

Crop failure can occur for many reasons. Pests; diseases; over-watering causing seed rotting and damping off of young plants; under-watering; and occasionally, but infrequently, poor quality seeds, insufficient preparation of soil, and so on. But more often than not it is caused by the vagaries of the British climate – so much so that even the most experienced professional horticulturist is not exempt.

In the summer of 1976 there was virtually a potato famine because of the lack of rain, and there were very few runner beans – even if they were thoroughly watered daily, the damage being caused by bees and ladybirds eating too deeply into the blooms, in search of moisture.

In the summer of 1977 there was an extremely poor apple crop, and virtually only ten per cent of the normal crop of blackcurrants. Runner beans and French beans were excessively late and few outdoor tomatoes ripened, all because of very cold winds and nights with temperatures down almost to freezing point at the end of June. The bean crops, when they eventually set, did yield truly extensive crops. The cold winds, and therefore temperatures, and late frosts also kept bees in their hives where they did not do their normal work of pollinating the blossoms. Bumble bees were few and far between.

It is wrong for the ordinary amateur gardener to be too downhearted or self-condemning when he has a crop failure because it is more than likely not his fault. He will find, however, that the climate adversely affecting one crop has been ideal for other crops. Nature has its own way of making compensation.

Left: Cabbage 'Hispi'.

Fruit

Currants (Black, Red and White)

The black, red and white currants are hardy, deciduous, fruit-bearing shrubs that thrive in all parts of Britain.

The black currant carries the greater part of its crop on stems produced the previous year and pruning is designed to encourage this annual replacement growth. The average height of the shrubs is 5 ft (1.50 m). Planting distance should be 6 ft (1.80 m) in each direction.

Red and white currants fruit on short spurs which shoot from the old stems. Pruning therefore, unlike the black currant, should leave the old branches to produce the fruit. The shrubs have an average height of 5 ft (1.50 m) and a spread of 4 ft (1.20 m). Planting distance should be 4 ft.

All varieties require soil that retains moisture but is also well drained. Red currants, especially, do not like soils that dry out in the summer and they will certainly not tolerate bad drainage. It is advisable to plant all types in a sunny position that is protected from cold winds during the late autumn. Black currants should be planted 2 in. (5 cm) deeper than they were in the nursery.

In the spring, mulch around the roots with well-rotted compost. With red and white currants, pull out any suckers that grow up from below ground, thus retaining only the old growth. Firm down with the heel after frosts. During dry spells, water well – particularly black currants.

In March, feed black currants with sulphate of ammonia (1½ oz (40 g) per square yard/metre) spread evenly over the full area of root. Give all varieties a dressing of sulphate of potash (1 oz (25 g) per square yard/metre). Keep weeds in check by hand weeding or by the use of a paraquat ('Weedol') rather than by forking or raking, which might damage the shallow root structure.

Birds, particularly tits and finches, attack buds on red and white currants, and all birds attack fruit as it ripens. If enough land is available, it is sensible to erect a permanent cage where fruit can be given protection all the year round.

Pruning: *Black currants.* After planting, cut all shoots down to 2 in. (5 cm) above soil level to an outward pointing bud. This will encourage growth to rise from below ground level, and strong shoots should be produced during the first season. At the end of the growing season, weak shoots should be removed to leave the strong shoots to crop the following season. During the autumn, after cropping, about half of the shoots which have carried fruit should be pruned to ground level. This practice of removing old shoots which have fruited and leaving in the young shoots should be followed every year.

In time the old shoots darken in colour, eventually becoming black, and young shoots remain brown. Hard pruning while bushes are young, combined with feeding and mulching, maintains a high proportion of the young shoots which are needed for fruit bearing.

Red and white currants differ from black currants in their fruiting methods and are pruned in a different way.

They usually grow on a short leg 9 in. (22.5 cm) tall and the lateral branches are pruned to promote short fruiting laterals. The branches should be cut back by about half their length to an outward facing bud in the autumn, and laterals should be cut back hard to within two buds. As the bushes become older some branches should be cut out altogether and be replaced by new shoots which should have their laterals pruned hard to within two buds. Red and white currants are a little more difficult to grow than black currant.

Above: Black currant 'Ben Nevis'.

Pests: *Birds, aphids* and *black currant gall-mites.*

Diseases: *Mildew, grey mould* and *coral spot* all respond to treatment by insecticides. *Honey fungus* causes the death of plants.

Left: Red currant 'Minnesota'.

Gooseberries

Gooseberries come from a hardy, deciduous, fruit-bearing shrub which can be grown in all parts of Britain. As they are prone to a number of virus diseases, the bushes should be obtained from a reliable specialist grower. They require a well-drained but moisture-retaining soil. Clay soils that dry out in the summer are not conducive to good crops, and organic compost should be added to the soil before planting.

Gooseberries should be planted between November and March in a fairly protected position away from cold winds but where they will have some sunshine. Height and spread are usually 5 ft (1.50 m) and planting distances should be approximately 6 ft (1.80 m) in each direction. As most varieties will produce good crops from May to August it is advisable to extend the season to its limit by planting early, middle and late varieties. Some varieties are especially suitable for dessert and others for culinary purposes, and some are suitable for both.

In the spring, apply a mulch of well-rotted compost around the roots to keep the soil moist during the following months. In late autumn, feed with sulphate of potash (1 oz (25 g) per square yard/metre) and in the spring apply sulphate of ammonia in the same strength. Because of the bushes' shallow roots it is best to control weeds by mulching rather than by digging or hoeing, but it is quite safe to use paraquat ('Weedol').

Late frost can cause a poor crop or even no crop at all. Birds cause damage to bushes, especially in the winter – protection by netting is a good investment. If there is enough land, it is wise to give gooseberries a permanent fruit cage in which they, and other fruits, are protected all the year round.

Pruning is important. The new bushes, if not already pruned by the grower, should be left until buds burst in the spring. When pruning takes place, build up the shape of the bush, and when that is done maintain a supply of young growth that will produce fruit. Do leave sufficient room for harvesting, and remember the prickly nature of the stems. Once the bushes are really established, hard pruning will improve the size of the fruit and support the vigour of the bush.

Pests: *Birds*, especially tits and finches, eat the buds in the winter. *Sawfly larvae* attack the leaves in early summer.

Diseases: *Mildew, cup rust* and *grey mould*, all of which are fairly prevalent, can be cured by appropriate insecticides.

Gooseberry 'Careless'.

Facing page: Gooseberry 'Leveller'.

Raspberries

This hardy, deciduous cane fruit, which does well in all parts of Britain, grows best in fertile, well-drained soil, particularly in a position sheltered from the wind, and partial shade is acceptable. The varieties are divided into two main groups – summer, which fruits in July and August, and autumn, which fruits in mid-September.

Raspberry canes should be planted during November in a soil rich in humus and well drained. They prefer an acid soil and will not thrive in alkaline soils unless they are given large quantities of organic matter. Because of the diseases that raspberries are subject to, it is important to buy canes from a reliable specialist grower.

Canes should be planted in rows 18 in. (45 cm) apart, along which wires should be run on 8 ft (2.45 m) high wooden posts that are sunk 2 ft (60 cm) in the ground. Three separate wires are necessary at 2 ft, 4 ft and 6 ft (60 cm, 120 cm and 180 cm) height from the ground, for the canes to be attached to as they grow. If more than one row of canes is grown, a distance of 4 ft (120 cm) is required between rows.

Moisture is necessary in the summer, especially while fruits are forming and ripening, and watering is advisable in dry weather. In the spring, apply a mulch of compost to the roots to retain moisture.

Feed with sulphate of potash (1 oz (25 g) per square yard/metre) in the winter and with half of that amount of sulphate of ammonia annually in March. Weeds should be controlled between rows by light hoeing, but care must be taken not to damage the shallow roots.

In the autumn, lift young new growth, retaining as much root as possible. Keep only the healthiest of young plants and replant them in fruiting positions cut back to a good bud 6 in. to 9 in. (15–22.5 cm) from the ground. Do not allow this young growth to fruit in its first year.

Left: Raspberry 'Glencova'.

Above: Raspberry 'Admiral'.

Pruning is necessary, except at the end of the first season when all strong canes should be left at full length and weak canes should be eliminated. When established, fruiting canes should be cut out and new shoots thinned out to leave five strong shoots to each plant. Once the canes are properly established they will spread rapidly and grow into spaces between rows. This growth should be checked by pulling up the unwanted canes while they are small to avoid too much disturbance to the root system. Autumn fruiting varieties should be pruned to ground level in February.

Pests: *Birds* of all kinds attack ripening fruit. *Aphids* and *raspberry beetle maggots* tunnel into the ripening fruit and make it inedible. Spray with insecticides.

Diseases: *Cane blight, cane spot chlorosis, crown gall, grey mould, honey fungus* and various virus diseases affect raspberries.

Rhubarb

For rhubarb the soil requires the addition of liberal quantities of manure, peat, charcoal or wood ashes and bonemeal, in a position in the garden protected from north and east winds. If this sounds like a lot of unnecessary trouble, remember that rhubarb is a permanent crop which justifies considerable care being taken in the preparation of the bed. It is best to start by buying roots from the nurseryman or garden centre, if possible in November but not later than March.

Rhubarb can be raised from seed, but it is a lengthy process and certainly not worth the bother when young roots can be obtained so easily.

When planting the roots, dig a hole sufficiently large to take the roots comfortably so that no parts need to be bent, and gently smooth them out over loosened soil. Cover the crowns with 2 in. (5 cm) of fine soil, tread down to firm, rake the surface lightly and cover with a 1 in. (2.5 cm) layer of manure. These young roots should be given a full year to establish themselves, therefore it is unwise to pull any sticks in their first season. When roots are established, harvesting will last up to three months but do not overpull, always leave some sticks to die down.

Remove all *flower* stems as they appear and, when pulling ceases, give a good mulching with more manure, preferably manure containing a good quantity of straw.

For wine, grow a separate crop from that grown for eating and leave the sticks until July. For jam, leave the sticks until the leaves begin to turn yellow.

Every third year, the roots should be lifted and divided. It is necessary to dig fairly deep all round the root to facilitate lifting without damage to the root. Remove the central crown and cut each root into two, leaving two crowns on each. Replant with the addition of a generous quantity of manure.

Forcing rhubarb is simple to achieve by covering the crowns in December with large pots or other suitable protection from the cold weather.

When harvesting, grasp the stem low down and pull away from the crown and the stick should come away easily with its basal bud. Always leave at least three sticks on each plant and do not pull sticks with unopened leaves. Harvesting should cease at the end of July.

Pests: *Eelworm* may infest plants and cause rotting. *Swift moth caterpillars* tunnel into the roots.

Diseases: *Crown rot* causes brownish rotting of terminal buds; may be caused by *late frosts*; *Honey fungus* will kill plants; *Leaf spot* causes irregular brown spots on the leaves.

Above: Forced rhubarb.

Left: Rhubarb 'Timperley Early'.

Strawberries

Strawberries, a favourite with most people, are not a difficult crop to grow. The plants grow best in soils rich in humus and, in consequence, the bed must be well prepared in advance. In all but the richest of soils large quantities of well-rotted compost or manure (up to 15 lb (6.750 kg) per square yard/metre) should be added, but because strawberries are not very deep rooted the humus should not be dug in too deep. If a rotovator is used, the compost may be put on the top of the soil and rotovated into the soil to a depth of 6 in. (15 cm). This will result in a fine rich mixture to a sufficient depth to have really healthy plants.

The next thing to do is to select a well-known and reliable specialist grower from whom to buy your plants. All growers have explanatory catalogues giving details of each variety in which they specialise – for size, for flavour, for heavy cropping, early crops, mid-season crops, September crops, those varieties most suitable for growing outdoors, in greenhouses, or in barrels, and so on.

Planting time is late July or early August for cropping the following year. If planting is not done until October, twenty-one months will pass before crops can be enjoyed. Those planted in late summer will require regular watering until the rains come.

Plants should be set 18 in. (45 cm) apart in rows 2½ ft (76 cm) apart and, provided sufficient compost or manure was dug in at the time of preparing the bed, no feeding with fertiliser will be necessary. Unless this practice is followed, too much leaf will grow and there will be very few strawberries. If, however, insufficient manure was used, it may be necessary to apply sulphate of ammonia in the spring (½ oz (12.5 g) per square yard/metre), or give a dressing of an all-purpose fertiliser (4 oz (100 g) per square yard/metre).

Facing page and above: Strawberry 'Red Gauntlett'.

The remaining work involved is to hoe regularly to keep down weeds. Cover with cloches from February onwards for early cropping, but only cover a part of the row to ensure a longer cropping period.

After frosts are over, cover the ground with straw or black polythene sheets which will keep the fruit clean and out of the soil, and the soil moist. If polythene is used, make small holes in the sheet and pull the leaves of the plant through and firm the polythene down around the roots. The extreme edges of the polythene should be covered with soil to avoid the wind from disturbing it once it is in position. Polythene also assists in keeping weeds at bay.

Dose well with slug bait and cover with netting to protect against birds.

Cultivation in Barrels

Barrels are available from garden shops and garden centres, ready bored with 2 in. (5 cm) wide holes 9 in. (22.5 cm) apart. The barrels should be filled with John Innes or Levington compost, putting plenty of crocks at the base for drainage. It is wise also to build an area of crocks through the centre to within 6 in. (15 cm) of the top to do the same job.

When the crocks are in position, lay turves of grass upside down on them and then the compost, firming it down well as you fill. As each hole is reached, insert a plant, making sure that the roots are well into the compost and its leaves are comfortably through the hole. At the top finish off with more plants at 9 in. (22.5 cm) intervals.

Keep the soil moist as the plants grow and give an occasional dose of liquid fertiliser.

Cultivation on Heated Greenhouse

Young plants should be potted in 5–in. or 6–in. pots at the end of July or in early August, in a good potting compost. Set plants to one side to assist the fruit when formed to hang over the side of the pot.

Allow the pots to stand in the open and to dry out gradually before

bringing them into the greenhouse, in November or according to the climate that season. In January restart them into growth again by watering and feeding once with a small dose of liquid fertiliser, gradually raising the temperature to 50–55°F (10–13°C) at night.

In March, when the flowers open and fruit is formed and ripening, raise the temperature to 75°F (24°C).

The flowers will need assistance in pollination or fruit will not set regularly. A camel-hair brush is recommended.

Water with care and give a liquid feed as fruit forms and begins to swell.

Pests: *Mice, slugs, strawberry beetles* and particularly *birds* eat ripening fruits. *Aphids, eelworms* and *mites* can infest plants, checking growth and causing leaf and flower disorders. In the greenhouse, *red spider mites* may cause trouble.

Diseases: *Grey mould* rots the fruit; *leaf spot* is due to fungi; *mildew* causes leaves to turn purple and curl; *virus diseases* cause leaves to crinkle and other leaf distortions.

A vegetable garden needs a deep freeze

No matter how careful the gardener, there are bound to be surpluses of almost any crop grown. Sell them, by all means, but also go a step further and keep some in a deep freeze for eating during the winter.

A freezer is an expensive piece of kitchen equipment. Therefore it is wise to buy a well-known and reliable make.

Apart from the small cubicle that is part of a refrigerator, and keeps vegetables frozen for only a limited time, there are two main types. The 'chest' variety, which is less expensive to run but rather difficult to manage, particularly when full. This type requires the keeping of a stock list which includes the latest date when each item must be eaten. In this type of freezer it can sometimes be difficult to find a particular item among all the numerous bundles of other foods. Some plastics manufacturers are now selling polythene bags in various colours which can be a great help in distinguishing the different produce.

The 'upright' variety, with its number of shelves at eye level and below, is more expensive to run but very much easier to manage because more or less everything can be readily seen and the stock is well split up by the shelves. Regrettably, this type is usually more expensive to buy than the chest variety.

The size to be chosen will largely depend on the space available in the kitchen, or perhaps garage, the size of the vegetable garden, and the number of people in the family. As a rough guide, 1 cu ft (30^3 cm) will store 25 lbs (11.250 kg) and 2 cu ft (60^3 cm) should be allowed for each member of the family.

When the annual maintenance service is being carried out, take all produce from the freezer – wrapped in a blanket it should keep perfectly satisfactorily for several hours. The same procedure is required when the freezer is defrosted every six months.

If a bag or box breaks in the freezer, it should be removed immediately to avoid contaminating other food. Every pack should be labelled and dated before it is put into the freezer.

Whole and sliced beans. SUTTONS SEEDS LIMITED

How to freeze your home-grown vegetables

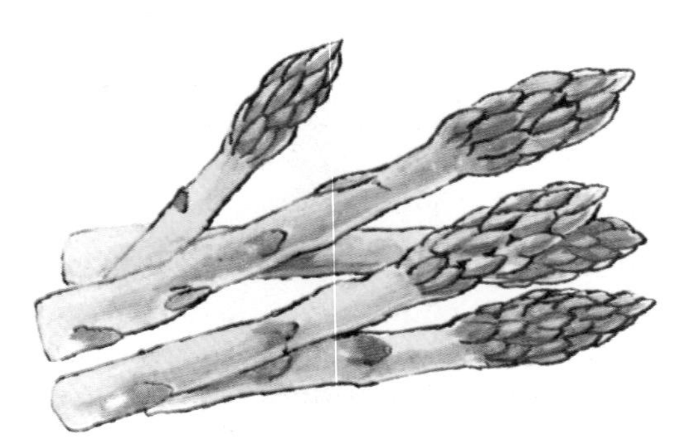

Asparagus
Freezing: Cut off the hard base end, sort into even lengths and wash well. Grade into thick and thin stems for ease of blanching and cooking. Blanch the thin stems for two minutes and the thick stems for four minutes. Cool in iced water, then drain. Pack in rigid containers, or wrap in small bundles, according to size, in tin foil.
Cooking: Place the frozen asparagus in boiling water, salted to taste, and allow to boil for five to ten minutes according to the size of the shoots.

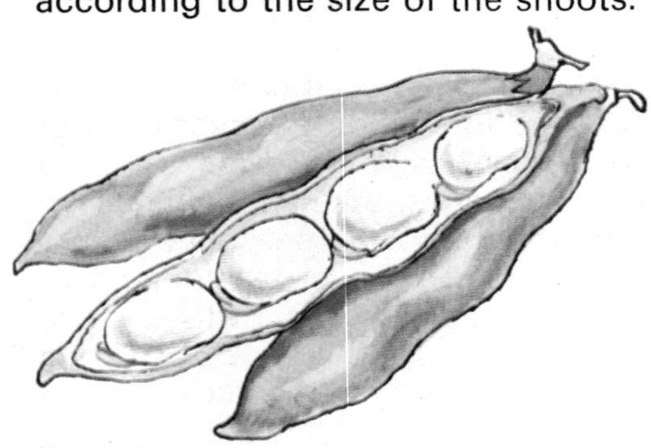

Broad beans
Freezing: Shell beans from the pods and wash well. Blanch for three minutes, cool in iced water, drain and pack in polythene bags.
Cooking: Place the frozen beans in boiling water, salted to taste, and allow to boil for eight minutes.

French beans
Freezing: Wash well and trim each end. Blanch for three minutes, cool in iced water, drain and pack in polythene bags.
Cooking: Place the frozen beans in boiling water, salted to taste, and boil for eight minutes.

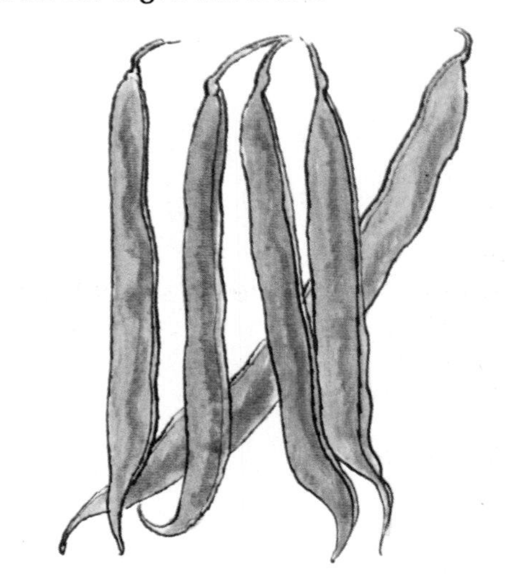

Runner beans
Freezing: Wash well and slice each bean into five or six slices, preferably using a slicer. Blanch for two minutes, cool in iced water, drain and pack in polythene bags.
Cooking: Place the frozen beans in boiling water, salted to taste, and boil for five minutes.

Broccoli
Freezing: Take off the outer leaves, wash well in salt water, divide into small spears and grade in sizes. Blanch the smaller spears for three minutes and the larger ones for five minutes. Cool in iced water and drain. Pack in rigid containers or wrap in small bundles according to size in tin foil.
Cooking: Place in boiling water, salted to taste, and allow to boil for six to eight minutes according to the size of the spears.

Brussels sprouts
Freezing: Use smaller, tightly packed sprouts. Remove the outer leaves and wash carefully in salt water. Grade into sizes. Blanch for two or three minutes according to size. Cool in iced water, drain and pack in polythene bags.
Cooking: Place the frozen sprouts in boiling water, salted to taste, and boil for eight minutes.

Cabbage
Freezing: Select young, fresh cabbages, wash them thoroughly in salt water, shred finely and wash again. Blanch for two minutes, cool in iced water and drain. Pack in polythene bags.
Cooking: Place the cabbage in boiling water, salted to taste, and boil for nine minutes.

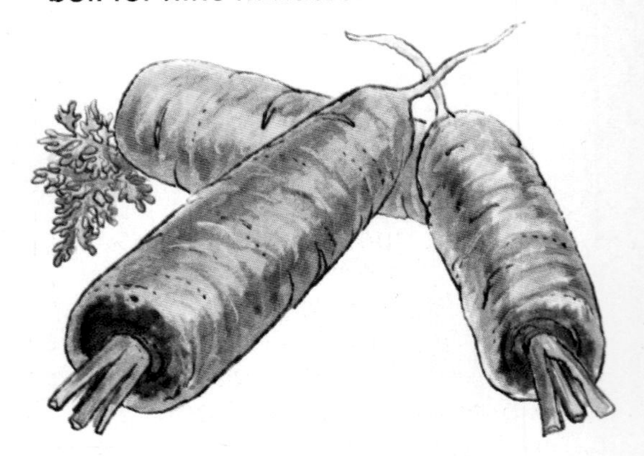

Carrots
This vegetable can be equally well

stored in sand in an unexposed position in the garden.
Freezing: Choose young carrots, trim their tops and tails, wash, scrape and leave whole. Blanch for four to six minutes according to size. Cool in iced water, drain and pack in polythene bags.
Cooking: Place the frozen carrots in boiling water, salted to taste, and allow to boil for eight to ten minutes according to size.

Cauliflower
Freezing: Choose white cauliflower, which must be both firm and compact. Wash well in salted water and break into florets; discard the main stalk. Blanch for three minutes, adding two teaspoons of lemon juice or vinegar to prevent discoloration. Cool in iced water, drain and pack in polythene bags.
Cooking: Place the frozen florets in boiling water, salted to taste, and allow to boil for eight to ten minutes.

Celery
Freezing: Choose young stalks, remove the main stem base, cut into short lengths 1–2 in. (2.5–5 cm) long. Blanch for three minutes, cool in iced water and drain. Pack in polythene bags or rigid containers.
Cooking: Thaw for 1–1½ hours before adding to casseroles or stews.

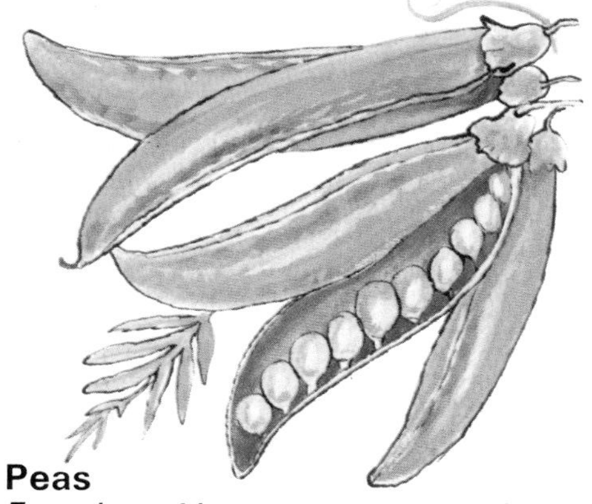

Peas
Freezing: Young peas should be selected, and podded and washed in salt water. Blanch for 1–1½ minutes. Cool in iced water, drain and pack in polythene bags.
Cooking: Place the frozen peas in boiling water, salted to taste. Add mint and allow to boil for seven minutes.

Spinach
Freezing: Select young, clean leaves and remove the stalks. Wash thoroughly in salted water and blanch for two minutes. Cool in iced water, and press out excess water. Pack in rigid containers, leaving ¾ in. (20 mm) headspace.
Cooking: Place the frozen spinach in boiling water, salted to taste, and allow to boil for six minutes. Frozen spinach may also be cooked in melted butter for seven minutes.

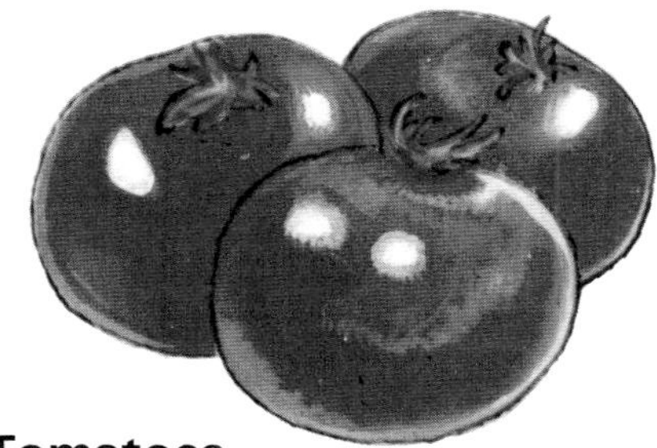

Tomatoes
Freezing: These may be frozen whole, or as puree, or as juice.

Whole: Place the tomatoes in boiling water, then skin. When skinned, pack in rigid containers or polythene bags.

Puree: Skin, core and simmer the tomatoes in their own juice until soft. Pass through a fine sieve or use a liquidiser, cool and pack in rigid containers. The puree is excellent for sauces, soups and stews.

Juice: Wash, cut in four and core the tomatoes. Simmer for five to ten minutes. Press through a very fine sieve, add salt to taste, cool, and pack into rigid containers. When required, the juice should be gently thawed in a refrigerator and served chilled with a little Worcestershire sauce added.
Cooking: The whole tomato is suitable for salads, and it is delicious when simmered in a saucepan with a little butter and eaten with cottage pie, sausages, liver and bacon, and similar dishes.

The ordinary refrigerator can also be used for very brief storage – just for a few days. Produce such as lettuce, beans, sprouts, cauliflower, sprouting broccoli, peas and carrots should be placed in a sealed polythene bag and put into the bottom shelf (crisper) where they will keep reasonably well for some days, in some cases for a full week.

Herbs

There is no doubt that the cook who has a good, working knowledge of herbs and their flavours will enjoy a reputation for producing attractive, enjoyable dishes. Moreover, the ability to make piquant sauces and savoury garnishes will help in sensible budgeting in removing the need always to buy the most expensive cuts of meat.

The wise gardener who decides to grow herbs will ensure that his herb garden is readily accessible to the kitchen and that its approach is not muddy. The reason is simple: more often than not herbs are picked in the middle of cooking, or preparing, a meal, and the person who is most likely to pick them is the member of the household who is cooking.

Most herbs originate outside Britain, in much warmer countries. The herb garden, therefore, needs also to be in a sheltered, warm position where it will enjoy as much sunshine as there is. If it is not possible to have an ordinary garden bed near enough to the kitchen and in sunshine, consider the possibility of constructing a raised bed – see under Asparagus on page 24. As an auxiliary supply that would be very much to hand, consider, too, having a window box outside the kitchen window.

Soil: In the main, herbs prefer a light, well-drained fertile soil, but they will in fact do almost as well in any soil that is properly prepared.

The site should be well dug in the autumn. If the soil is heavy add compost or peat giving approximately 2 bucketsful per square yard/metre. During the winter give the bed a fair dressing of lime, as all herbs except sorrel prefer an alkaline soil. In March, or as soon as the soil allows, rake until a fine tilth has been obtained. Finally, rake in a dressing of all-purpose fertiliser.

Sowing: Now is the time to decide which herbs are to be grown. A selection that includes borage, chives, fennel, mint, parsley, rosemary, sage, sorrel, savory, tarragon and thyme, with perhaps a bay tree growing elsewhere in the garden, will be sufficient for most people.

Hardy annuals may be sown in March, and the half-hardy herbs at the end of May. Perennials should be planted either in April or in the autumn when digging has been completed. It is likely that only one or two plants of most herbs will be required, and in that case plants may be obtained from a garden centre. If larger quantities are required – parsley is an example – seeds should be sown in drills ½ in. (12.5 mm) deep and covered with sifted soil firmed down with the back of a rake.

The following herbs, many of which are already well known and often used, will, if used properly with skill, add enormously to the pleasure to be found in well-cooked food.

Anise

This is slow growing and flowers three months after planting. It grows to a height of 18–24 in. (45–60 cm). The lower leaves are oval with serrated edges; the upper leaves are longer and divided into three segments. The flowers are yellowish-white.

How to grow: Anise does best in a well-dug soil because of its long tap root. It requires a sunny position where plants should be thinned out to 6 in. (15 cm) apart in rows 2 ft (60 cm) apart.

Harvesting: The seeds should be harvested about one month after the flowers bloom.

How to use: Anise is famous for its liquorice flavour and is used specially in making bread and rolls.

Basil

Basil is a spicy herb with a slight flavour of pepper and is grown for the important seasoning it adds to many dishes.

How to grow: Basil is easy to grow. Sow the seeds in May in a sunny position. As they grow, pinch out the stem tips to encourage strong and compact growth. Do not apply fertilisers as lush growth will reduce the flavour.

Harvesting: Gather the leaves when flowering begins. They may be preserved by freezing.

How to use: Basil is excellent for flavouring soups, stews and sauces, Shredded leaves give a piquancy to green salads.

Borage

This herb grows 2–3 ft (60–90 cm) tall. It has grey-green leaves and flowers that are bright blue and star shaped.

How to grow: Borage likes any fertile soil but does not transplant easily. Sow the seeds in September or April in a sunny position; or, to lengthen the harvest sow once in the autumn and twice in the spring – in April and May.

Facing page: Rosemary.
ARDEA LONDON

Harvesting: Pick the leaves before the flowers bloom.

How to use: Chop the leaves – which are cucumber flavoured – and add to green salads.

Caraway

Caraway is a biennial and the seeds are produced in the second year. It grows to 2 ft (60 cm) high and has attractive clusters of greenish-white flowers. Seeds ripen a month after flowering: the plant then dies.

How to grow: Sow seeds in the autumn and again in the spring. Thin out to 12 in. (30 cm) apart. Some protection might be needed in very cold weather.

Harvesting: Collect the seeds when they ripen, about a month after flowering.

How to use: The sharp, aromatic seeds are used in cakes, rolls and bread as a seasoning. The oil from the seeds is used for flavouring the famous liqueur Kummel.

Chervil

This herb looks well in a flower border with its light green fern-like leaves similar to parsley. It grows in height to 12–18 in. (30–45 cm).

How to grow: Sow seeds at intervals from March to August and thin out to 12 in. (30 cm) apart. A supply for the winter can be obtained by sowing under glass from October onwards. The plants prefer partial shade and moist soil. To improve the foliage, cut the flower stems before they bloom.

Harvesting: Pick the leaves before the buds break. The green tender leaves may be dried.

How to use: Chervil has a mild parsley flavour and is excellent for sauces. It adds a delicate flavour if sprinkled on such dishes as chicken and fish.

Chives

The tubular mid-green stems, with an onion flavour, grow to about 6 in. (15 cm) long. The purple flowers should be cut immediately they appear.

How to grow: Chives thrive in any well-drained soil, preferably in a sunny position. Plant roots at any time from September onwards. Keep moist at all times. Dig up and separate every three years.

Harvesting: Pick the stems as they grow, nipping them off as near the root as possible. Harvesting should continue from April until the first frost.

How to use: Cut the raw stems into short lengths and add to salad dressings, potatoes and dishes where an onion flavour is required.

Dill

Dill is an annual that grows to a height of 2–3 ft (60–90 cm) and bears lacy, light green leaves. The tiny leaves produce large quantities of seed.

How to grow: Sow in a sunny position in well-drained soil. When fully grown, plants will need staking.

Harvesting: Pick leaves as soon as the flowers begin to bloom. Gather the seeds when they are flat and brown in colour.

How to use: Dill is notable for both its seeds and leaves. The seeds have a slightly bitter taste but the leaves have a delicate bouquet. The leaves are excellent sprinkled over many dishes, including meat, fish and vegetables, and are pleasant added to green salads. The seeds are used largely for flavouring bread and rolls, and in some cheeses. They are also a useful addition to pickles and beetroot dishes.

Fennel

This is a hardy perennial which forms clumps with sweetly aromatic blue-green leaves. Numerous 3 in. (7.5 cm) wide flowers appear in July.

How to grow: Sow seeds in March in a sunny position that has well-drained soil. Thin out to 24 in. (60 cm) apart to allow for full growth, which can be more than 5 ft (1.50 m). Staking is necessary.

Harvesting: Pick the leaves just before the blooms appear.

How to use: Fresh or dried leaves may be used for herbal tea and for flavouring sauces. The herb is excellent with white meat such as chicken, pork and veal. Chopped leaves add an aniseed flavour to soups. The seeds go well with gherkins and cucumbers.

Garlic

Garlic is a strong onion-like perennial with grassy grey-green leaves.

How to grow: Plant individual cloves or bulblets just below soil level 6–9 in. (15–22.5 cm) apart. They will do best in a light, well-drained soil in a very sunny position. Remove the rounded heads of the white starry flowers as they appear during June.

Harvesting: When the foliage turns yellow, usually in August, lift the bulbs and allow them to dry thoroughly in the sun. Tie the dried bulbs into bundles and store in a dry, airy place, free from frost, ready for use when required.

How to use: Garlic cloves may be used to season a wide range of dishes, but they should be used sparingly because of their strong flavour.

Marjoram

This herb is a half-hardy annual that grows to a height of 2 ft (60 cm) and is similar in flavour to thyme.

How to grow: Sow seeds under glass in March, prick out into $2\frac{1}{2}$–in. pots and, after hardening off, into the herb garden in May in a sunny position.

Harvesting: The sweetly spiced leaves may be picked at any time, but they are most flavoursome just before the blooms appear. They dry very well.

How to use: The leaves have a slight flavour of mint and go well with many vegetables. They are also pleasant when sprinkled on omelettes, scrambled eggs and some soups.

Mint

There are a number of varieties which include spearmint, orange mint, peppermint and apple mint. The common mint, or spearmint, is a vigorous hardy perennial.

How to grow: Mint can be grown from seed, but it is best to obtain the first root from a garden centre. The root will grow well in either sun or shade and its underground growth will spread rapidly. Dig up and separate the roots and replant every four years.

Harvesting: The more frequently the stems are cut, the better the growth.

How to use: Chopped leaves are used for garnishing meat, potatoes and carrots, and when mixed with vinegar as mint sauce with lamb.

Parsley

Almost everybody knows parsley, both as a herb and as an attractive border plant. There are a number of varieties to choose from, but all are very slow to germinate.

How to grow: Sow seeds in March and again in July to ensure an all-the-year-round crop. Germination usually takes eight weeks, but soaking the seeds in tepid water for twenty-four hours does help. Thin out 4–6 in. (10–15 cm) apart.

Harvesting: Pick the largest sprays as they grow and do not allow any to seed.

How to use: Pick the leaves from the stalks and chop them into small pieces. Use them to garnish meats, fish, salads and potatoes. To dress asparagus, artichokes, carrots and new potatoes, heat the chopped leaves in butter.

Rosemary

Rosemary is a bushy hardy evergreen shrub about 5 ft (1.50 m) tall with grey-green foliage and pale blue flowers.

How to grow: Plant young rooted cuttings from a nursery in a sunny position. Growth must be controlled by frequent pinching of the tips.

Harvesting: Simply pick small sprigs of leaves.

How to use: Rub the rosemary into pork, veal and lamb roasts, or sprinkle on dishes of chicken, rabbit, hare and grilled and fried fish.

Sage

This is a hardy evergreen shrub that grows to a height of 2 ft (60 cm) with oblong wrinkled grey-green leaves and purple flowers.

How to grow: Start from seeds or, better still, cuttings which when rooted will require a sunny position. Space plants 2 ft (60 cm)

apart. After three years plants become woody and need replacement.

Harvesting: Pick the leaves before the blooms appear and cut back the stems when blooming has finished.

How to use: This aromatic herb is notable as a stuffing herb and is often mixed with onions for this purpose. It is excellent with duck, chicken, veal, pork chops, turkey, rabbit and hare. Sage is sometimes used in making herb bread.

Savory

This hardy annual has sweetly aromatic, narrow, dark green leaves, and spikes of purple or lilac tubular flowers that develop from July to September.

How to grow: Sow seeds in April and thin out to 6 in. (15 cm) apart. Savory requires sunshine but less watering than most herbs.

Harvesting: Gather leaves before the blooms appear.

How to use: Savory goes well with all kinds of beans, and with salads, soups and vegetable dishes. It is a mild herb and is excellent in egg dishes and cheese soufflé. It does well as a garnish for poultry and in stuffings.

Sorrel

Sorrel is a tufted hardy perennial which has fleshy triangular grey-green leaves. Small, pale, green-red tinted flowers develop from June to September.

How to grow: Although sorrel can be grown from seed it is better to buy a young rooted plant in the autumn. Thereafter, the crop can be increased by dividing the plant in the spring. It enjoys a sunny position and moist soil.

Harvesting: Simply pick the larger leaves.

How to use: The leaves are slightly bitter and are used in garden salads. They may also be used as an alternative to spinach and made into a purée with butter which is fine served with veal, pork, chicken and egg dishes. Finely chopped leaves improve many soups.

Tarragon

A perennial plant which may need some protection in very cold winters. It grows to a height of 2 ft (60 cm) and has fine, dark green leaves with pointed tips. Very small whitish-green flowers appear in August.

How to grow: The first plant should be bought from a nursery. The stock can be increased by taking cuttings or by dividing small pieces of root which should be potted and set out in the following spring. Tarragon prefers only partial sun and a fertile, well-drained soil.

Harvesting: Pick the leaves when young, before flowers appear. They may be dried or frozen.

How to use: Tarragon should be used with restraint with fish dishes, including shell fish. It also improves poultry, omelettes and green salads. Added to butter, it makes a dressing for asparagus and artichokes.

Thyme

Thyme is a shrub-like plant that grows to 8–12 in. (20–30 cm) high. It has slender, woody branches and tiny grey-green leaves about $\frac{1}{4}$ in. (6 mm) long. Purplish flowers grow in June. There are several varieties but the 'common garden' form is recognised as having the best flavour.

How to grow: Germination of seeds can be difficult, so it is best to start with a bought rooted plant and thereafter to propagate by cuttings or dividing. The plants enjoy maximum sunshine and a relatively dry soil. Keep the plants well clipped to prevent them becoming woody.

Harvesting: Clip the plants and use the leaves.

How to use: Thyme is a pungent herb and an essential part of 'bouquet garni'. It has a strong, warm, clover-like flavour. It mixes well with most other herbs and is excellent for stuffings and for garnishing carrots, onions and beetroot. Slow-cooking beef dishes are improved by its flavour.